T0133594

Deep Neural Network Applications

Hasmik Osipyan
Software Development Manager
Synopsys Inc., Yerevan, Armenia

Bosede Iyiade Edwards
Imagineering Institute, Nusajaya, Malaysia

Adrian David Cheok
Professor, Professional University of Information and Management for Innovation, iUniversity, Tokyo, Japan

CRC Press is an imprint of the
Taylor & Francis Group, an **informa** business
A SCIENCE PUBLISHERS BOOK

First edition published 2022
by CRC Press
6000 Broken Sound Parkway NW, Suite 300, Boca Raton, FL 33487-2742

and by CRC Press
2 Park Square, Milton Park, Abingdon, Oxon, OX14 4RN

CRC Press is an imprint of Taylor & Francis Group, LLC

Library of Congress Cataloging-in-Publication Data (applied for)

ISBN: 978-0-367-21146-2 (hbk)
ISBN: 978-1-032-22903-4 (pbk)
ISBN: 978-0-429-26568-6 (ebk)

DOI: 10.1201/9780429265686

Typeset in Times New Roman
by Radiant Productions

From Hasmik Osipyan to anyone who believes in technology and is not afraid of AI (especially to my mum Anahit)!

From Bosede Iyiade Edwards to the future; and to Boluwatife, Inioluwa, and Opeyemi, the future that I see today!

From Adrian David Cheok; for my sweet pea daughter Kotoko, I love you infinity percent forever and ever!

Foreword

After many crises, AI has entered the era of deep learning and neuro-networks, and has finally reached the stage of full-scale diffusion.

The 21st century, with three-quarters left, will be the age of AI and data.

AI will spread not only to entertainment and business such as games and self-driving cars, but also to public fields such as healthcare, global warming countermeasures, and security, and will become a social infrastructure.

A new understanding of technology is needed.

Learn the cutting edge of research.

Professor Ichiya Nakamura
President of iUniversity (Professional University of Information and Management for Innovation), Tokyo, Japan

Preface

Artificial intelligence is one of the hottest topics in the world today, and deep learning is a subset of machine learning in artificial intelligence. Deep learning is always fun with the latest features like autonomous vehicles and fraud detection. I could not imagine the possibilities today, but now I cannot even imagine a day without AI. So, this book goes into more detail on this remarkably interesting topic, “deep learning”.

Deep learning is the beginning of what a machine can do, and developers and business leaders need to understand what it is and how it works. Deep learning models are adept enough to focus on the exact function itself, requiring some instruction from the programmer and helping solve dimensional problems. Therefore, deep learning algorithms are used, especially if there are a lot of inputs and outputs.

Now let us take a closer look at the various aspects of deep learning.

Why is it called Deep Learning?

Deep learning is the work of artificial intelligence that reflects the activity of the human brain by preparing information and creating signals for use in decision making. Also known as deep neural learning or deep neural networks. This is a type of machine learning that prepares a computer to perform human-like tasks such as audio observation, image identification, and prediction.

Deep learning is computer specific by setting essential boundaries on information and observing design using multiple layers of processing, rather than organizing information to overcome predefined conditions. The network trains to learn.

Deep learning has networks that deserve unsupervised learning from unstructured or unlabelled information. Put simply, deep learning is a type of algorithm that clearly works well for predicting things. Deep learning involves training and learning data from previous experiences. This is a promising way to make computer education imaginable. This machine learning mimics a human neural network.

This artificial neural network saves us time because it prevents people from conducting monotonous activities and this procedure reduces the risk of human error and therefore has a further total compliance.

Benefits

The deep learning architecture is flexible to adapt to unfamiliar problems in the future. New features can be developed from a convenient and limited training data set.

Disadvantages

Training is awfully expensive due to the complexity of the data model. It is difficult for low-skilled people to adopt because there is no basic theory that teaches how to choose the right deep learning tool.

How Does it Work?

Deep learning works on the concept of repetitive teaching. Train your computer to understand specific patterns and identify images and sounds. Once recognized, the computer can automatically pick up the word or voice.

This learning method is not vastly different from our human learning method. As a child I also learned to listen to the things around me and to say the words. This is the method we have learned, and deep learning now uses this formula to teach computers.

The Importance of Deep Learning

Deep learning is especially important because it makes your tasks precise and fast. The ability to manage large amounts of detail makes deep learning extremely robust when working with undeveloped data. Therefore, computer vision is a splendid example of an activity that makes deep learning logical for business applications. Deep learning allows you to identify your face on the Internet.

This book provides everyone with a deep understanding of the topic of neural networks. We hope everyone can use this book to make their own exciting prototypes and products.

Contents

Foreword **v**

Preface **vii**

1. Introduction **1**

1.1 Artificial and Deep Neural Networks 1
1.2 Evolution: Where are We Now? 4
1.3 Motivation 8

2. Deep Learning Basics **11**

2.1 Applied Math 11
2.1.1 Forward Propagation 11
2.1.2 Backward Propagation 14
2.1.3 Optimization of DNNs 17

3. Neural Network Structures **29**

3.1 Feed Forward Neural Network 30
3.2 Convolutional Neural Networks (CNN) 33
3.3 Deconvolutional Neural Network (DNN) 37
3.4 Deep Convolutional Inverse Graphics Network (DCIGN) 37
3.5 Recurrent Neural Network (RNN) 39
3.6 Kohonen Network (KN) 40
3.7 Deep Residual Network (DRN) 42
3.8 Long Short-Term Memory (LSTM) 43
3.9 Gated Recurrent Neural Networks (GRU) 46

3.10 Bidirectional Recurrent Neural Network (BRNN) 48
3.11 Hopfield Network (HN) 48
3.12 Generative Adversarial Networks (GAN) 50
3.13 Deep Belief Network (DBN) 52

4. Top Applications of Deep Learning Across Industries **57**

4.1 Agriculture 57
4.2 Banking and Finance 62
4.3 Education 66
4.4 Healthcare 71
4.5 Legal and Politics 76
4.6 Military and Security 79
4.7 Service and Marketing 85
4.8 Social Media and Entertainment 91
4.9 Transportation 100
4.10 Other Applications 106

5. Discussions and Criticism **111**

5.1 Challenges We Face Today 111
5.2 Future of Deep Neural Networks 114

References **119**

Index **145**

Chapter 1

Introduction

1.1 Artificial and Deep Neural Networks

Artificial Intelligence (AI) is everywhere at the moment, including virtual assistants, image recognition systems, search recommendation systems, and other applications. And there is a good chance that you are probably using one of the top AI applications during your daily life. Consequently, AI is a powerful and interesting tool that is getting more and more attention day by day.

There are many terms associated with AI, such as Machine Learning (ML), Neural Networks (NN), etc. So, we can think of AI as a super class that includes all of these subsets, and with the main purpose of making computer systems perform actions which will be considered intelligent [208, 194]. However, it is difficult to give a definition for computational intelligence because it is strongly associated with that period of time. While technology advances, the previous benchmarks that defined the term of intelligence become outdated. For example, the action of recognizing text through optical character recognition was described as intelligent a long time ago. This functionality is no longer considered to be intelligent for machines at this time, as this function is now taken for granted as a general feature for computing systems. Consequently, the demand for computer intelligence and AI is changing every year. Currently, the goals of AI include reasoning, learning, planning, perception, knowledge representation and creativity. There is already a lot of research work going on towards creating computing systems that will achieve these goals [45, 148, 146, 21, 192]. However, there is still a long way to go, and AI will continue to act as a technological innovator for the foreseeable future.

One of the more popular terms in recent times is Deep Learning (DL), which describes certain types of NNs and related algorithms that often process very raw input data through many layers of nonlinear transformations [64]. The DL associated algorithms are used not only for supervised learning problems, but also for unsupervised and reinforcement learning problems. In supervised learning, the model learns from a supervised labeled data set with guidance, whereas in unsupervised learning, the machine learns from the unlabeled input data set without any guidance. Compared to these two learning methods, during reinforcement learning, the machine interacts with its environment and learns by trial and error method [175]. For example, DL also excels in the field of unsupervised feature extraction for automatically deriving or constructing meaningful features of the input data that will be used for further learning. To better understand DL concepts, we will first review the idea behind NNs.

Artificial Neural Networks (ANNs) are computing systems that try to replicate the idea of biological neural networks in animal brains [223, 76]. Different parts of the brain are responsible for processing various pieces of information, and these parts of the brain are arranged hierarchically, so called in layers. Therefore, ANNs are trying to simulate this layered approach of processing information and making decisions. Such computational models are learning and improving their knowledge based on the given examples. In its simplest form, an ANN can only have three layers of neurons. The input layer is designed for entering the data set into the system. Then, the hidden layer processes the information, while at the output layer, the system decides what to do based on the processed data. However, ANNs can get much more complex than that, and include several hidden layers. The part of the broader family of ANN that is made up of more than three layers (multiple hidden layers) is called a Deep Neural Network (DNN), and this is what lies at the heart of DL (Fig. 1.1). Different layers of DNN are required to extract different features until they can recognize what they are looking for.

The DL learning system is self-learning as it progresses by filtering information through multiple hidden layers, similar to the human brain [64]. The DNN model consists of artificial neurons (represented by circles in Fig. 1.1), which are organized into layers that perform various transformations on their inputs. The number of neurons in the first input layer is defined by the input data set. The input layer passes the inputs to the first hidden layer, that performs mathematical computations on the input data. One of the difficulties when creating DNNs is determining the number of hidden layers and the number of neurons for each layer. One thing is clear: whether it be three layers or more, information flows from one layer to another, just like in the human brain. The connection between neurons (synapses) can transfer a signal from one neuron to another. These signals travel from the first (input) layer to the last (output) one, usually after passing a hidden layer consisting of units that transform the input into something that the output layer can use Fig. 1.1. These are excellent machine learning tools for finding patterns which are far too complex for a programmer to extract and teach the machine to recognize. At the end, the output layer returns back the calculated output data. The basic math, the main concepts (including neurons, weights, biases, synapses and functions that try to replicate the human brains) and how DNNs calculate the final results will be thoroughly described in Chapter 2.

As brain-inspired systems designed to replicate the way humans learn, NNs modify their own code to find the connection between input and output. Hence, AI has benefited greatly from the arrival of NNs and DL. So, how does DNN work in general? Humans learn from their everyday experiences. The tasks they do over and over again gradually become more efficient and the percentage of errors/mistakes decreases. Following the same principle, the NNs also require data for training, and more data should be used in advance to get accurate results. In general, there are three main sets of data that need to be divided when working with NNs. First, this is the so-called training set, which helps NN establish various weights between its nodes. Second is a validation set required for fine tuning. Finally, the third is a test set to check if it can successfully turn input into desired output. Once this is done, the researchers who have trained the network can label the output, and then use back-propagation to correct any mistakes which have been made. After a while, the network will be able to carry out its own classification tasks without needing the help of researchers each time.

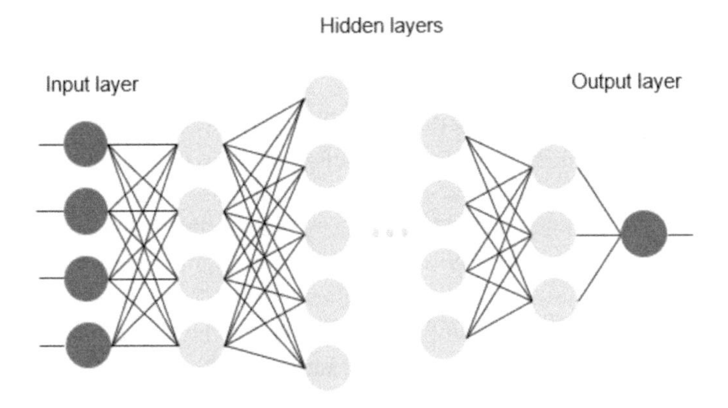

Figure 1.1: The structure of deep neural networks.

All of these NNs come with their own complexities and use cases, which we will discuss in Chapter 3. In addition, choosing the correct NN for a task depends on the data set and the specific application for which it will be used. There are still many challenges for NNs that researchers are working on to address. One of the biggest issues is training time, which increases with the complexity of the task. Usually the computation time is higher due to the many training parameters NNs required.The next challenge is that NNs are black boxes, where the user feeds in data and receives answers. Thus, users are not involved in the decision-making process, which complicates the situation. There are also some challenges on the technical level, such as overfitting, which is the result of multiple abstraction layers.

Nowadays, NNs typically have from a few thousand to several million units with millions of connections. However, this number is still several orders of magnitude less than the number of neurons in the human brain, which means that these networks can perform many tasks at a level beyond that of humans. Thus, there is still a lot of work to be done to achieve all the goals of AI and to actually simulate the human brain.

1.2 Evolution: Where are We Now?

When we think of AI or NNs, most of us refer to 21st century inventions. To all our surprise, the concept of NNs actually dates back to the 1940s [72]. One of the reasons most of us think it is a new invention is that in the early stages, the approaches and algorithms were unpopular due to their various shortcomings. However, it is important to note that some of the algorithms developed in those old times are still widely used today in different DL approaches.

In the early stages, researchers were trying to replicate human intelligence only for specific tasks, such as winning games. A bright example is the Deep Blue machine, developed by IBM to win a chess match against a world champion [25]. In such systems, in order to achieve better results in solving specific tasks, researchers introduced a number of rules that the computing system needed to follow. In the end, to win the game, the computing system made decisions based on already defined rules. At this early stage, advancements were made in the name of cybernetics based on the idea of biological learning [112]. Cybernetics has been started by the development of McCulloch-Pitts Neuron for simulating the biological neuron [133], followed by the Perceptron, developed by Frank Rosenblatt for learning weights automatically [169]. However, it has been proven that single-layer Perceptrons could not solve the seemingly simple XOR (exclusive OR) classification problem [135].

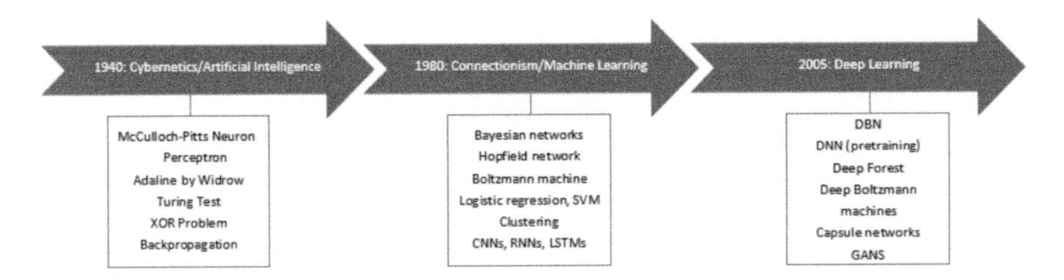

Figure 1.2: The timeline of AI, ML and DL.

In Fig. 1.2, we show a timeline of AI, ML and DL development, as well as the main inventions over these periods. AI research began as an academic discipline in its own right in 1956, inspired by the work of the renowned mathematician Alan Turing. In his seminal paper "Computing Machinery and Intelligence", which was published in the philosophical journal "Mind" in 1950, Turing proposed the possibility of thinking machines [198]. In the test, if a human evaluator cannot determine whether is talking to a machine or to a human based on text conversations, then the machine has passed the test. In 1956 a conference took place at Dartmouth College, in Hanover, New Hampshire – the Dartmouth Summer Research Project on Artificial Intelligence - where the term "artificial intelligence" emerged [112]. Later, the Alan Turing test became widely used to measure the intelligence of the very first chatbots, such as ELIZA [209], and only in 2014 the first machine, Eugene Goostman, passed the test [180]. In the period of cybernetics, as with the Turing test, there were many inventions that are still used in modern approaches. For example, back-propagation techniques or the Adaline learning function developed by Bernard Widrow, which is similar to the stochastic gradient descent, currently used in modern approaches [210].

Connectionism became popular in the early 1980s when Parallel Distributed Processing started to be widely used. It was a first movement in cognitive science that tries to understand how the human brain works at the neural level [19]. This was a period when ML and NNs attracted the attention of researchers. It referred to the ability of computing systems to learn their next actions using large data sets instead of hard-coded rules. Hence, ML has allowed computers to learn on their own and make decisions based on their past experience. Various models were developed such as Long Short-Term Memories (LSTM), Convolutional Neural Networks (CNN), Support Vector Machines (SVM) or Boltzmann machines, which still remain key components of different advanced applications of ML and DL [31]. However, during connectionism, all these methods did not get wide usage and did not achieve the promised results due to a lack of computational resources.

Later, in 2006 the third wave of DL started, and it became possible due to the processing power of modern computers that can easily process large sets of input data. Actually, the expression DL was first used when talking about ANNs by Igor Aizenberg and colleagues around 2000 [4]. We can think of DL as a method for training machines with thousands, or even millions of artificial neurons. Until now, many different approaches have been invented to solve the problems encountered when training traditional neural networks, such as slow learning or the requirement of a big training data set. One of this new inventions is Deep Belief Networks (DBN), introduced by Geoffrey Hinton and used Greedy Layer-wise Training [82], with simple implementations were Boltzmann machines. Generative Adversarial Networks (GAN) enable models to solve unsupervised learning, which is more or less the ultimate goal of the AI community [65]. The advancement of DNNs also helps to overcome the shortcomings of the initial problems that were associated with rule-based solutions. Previously, a researcher would teach a computing system the rules, rather the machine learning them on its own. This was a limited solution that required researchers to constantly intervene in the process, and it was solved through the development of DNN. For example, compared to Deep Blue, AlphaGO won the world championship in Go by training and learning itself on a large data set of expert movements [204]. Consequently, the DL breakthrough has completely changed the development of AI, becoming part of our daily life. The Facebook social network, Google search engine and Youtube recommendation engine all use DL.

Hence, aided by the arrival of DNNs and the algorithms from the Connectionism period, AI has started to truly live up to its potential today. Currently, this renewed interest in DL is mainly associated with factors below:

- Capabilities of modern computers,
- Access to open-source libraries such as Keras, TensorFlow, PyTorch [202],
- Available large data based on online services, which increases the accuracy for various models.

Although we are still in the early stages of AI and DL adoption, we have already seen significant impact across nearly every industry. Self-driving cars and human-like virtual assistants are just a few products of the recent shift in AI research that has revolutionized the way machines learn from data. In addition, it is clear that the further evolution of AI and DL will affect humanity in all areas. There are many contradictory opinions among researchers about whether millions of jobs will be lost due to advances in AI, and where AI will finally bring humanity to, including it will be a threat to humans, or help in making smarter decisions [170]. AI supporters believe that it will help us solve challenges such as various diseases, global warming, and hunger in the world. On the other hand, opponents fear that advances in technology will cause people to lose jobs and creativity as machines will become smarter than humans. While it is interesting and wise to look into the future to predict where the evolution of AI will take us, we need to remember that we are still in the early stages of our trip. The various constraints we still face in AI and DL show that we are yet far from machines that will think and behave like humans. As the 2018 year Turing Awardee Yann LeCun said: "Machines are still very, very stupid. The smartest AI systems today have less common sense than a house cat."

Based on capabilities there are currently three types of AI being developed: Narrow (week), Strong (general) and Super AIs [206]. While Narrow AI is focused for managing small data sets to solve specific tasks, Strong AI is designed for performing any kind of intellectual task like humans. Super AI is dedicated to surpassing human intelligence and abilities. The Narrow AI is the most common and currently available type in the world of AI. Apple Siri, AlphaGo, self-driving cars are good examples of Narrow AI. Currently, researchers all around the world are focused on developing systems with Strong AI. While Narrow and Strong AIs are extremely useful, both of them are not going to take over the world. Meanwhile, Super AI is likely to remain a subject of science fiction for a long time to come.

Without DL techniques and algorithms, it will be impossible to achieve Super AI vision. In their turn, existing DL architectures and algorithms are categorized into three classes, of which we will review thoroughly different approaches in Section 3.

- Generative modeling is unsupervised learning that automatically discovers patterns in the input data and then generates new design options based on the learned joint probability distribution.
- Discriminative architectures learn the conditional probability distribution and model the boundary between classes.
- Hybrid method combines the properties of the generative and discriminative architectures.

DL is now widely used in different applications across industries, and during this period, it has increased accuracy over other approaches. Based on the evolution and history of AI and DL, it is important to note that it took decades to achieve these results. Nowadays, more and more companies from different industries are starting to use these approaches to solve their day-to-day tasks. However, DL still has a long way to go!

1.3 Motivation

AI and related technologies are evolving day by day, and the applications for AI are becoming endless. Nowadays, deep neural architectures have acquired great achievements across a variety of industries including defense, healthcare, computer vision and more. Researchers have even been working on creating a software that borrows concepts from Darwinian evolution to build AI programs that improve themselves without human intervention. While decision making and human-like thinking by computational systems is still a long way off, there have already been significant advances in various applications based on ANNs and DNNs. However, even now, the solutions based on NNs for such techniques and applications are sometimes incredibly complicated. In addition, the big question remains: "What's next"?, for which there is no yet simple answer. Considering how vast the field of AI and its implementations could be, it is difficult to predict one field in which AI will excel in the future. However, there is no doubt that AI will be used in more situations in the coming years.

Currently, the main challenge with AI systems is that humans often do not know what goals to give for these computing systems, because sometimes we do not know what we really want to achieve. For example, ultrasafe self-driving cars drive too slowly and brake so often that they make passengers feel sick. This is understandable, because when researchers try to list all the goals and preferences that a self-driving car should simultaneously fulfill, the list inevitably becomes incomplete. In addition, allowing machines to define their own goals, so-called "autonomous" machines, can become increasingly risky. Due to huge available computing capacities, "autonomous" machines can become more intelligent, and they will ruthlessly pursue their reward function and try to prevent us from turning them off. Therefore, instead of creating Super AIs and letting machines pursue their own goals, they should seek to satisfy human preferences; their only purpose should be to learn more about preferences and real needs of humankind. That will help keep AI systems safe.

AI, powered by the most capable tools like DNNs and the more sophisticated DL techniques, is an extremely influential and exciting field. This will become more and more important in moving forward, and will certainly continue to have very significant impacts on all industry fields. Driven by the importance and relevance of DNNs and AI in general, in this book we offer new perspectives for the future exploration of DNNs and their applications in various domains, along with explanation of the approaches. We first review the mathematics behind NNs, then we focus on existing architectures and algorithms for top NN structures. Furthermore, to illustrate how industries are motivated to use the NN structures, existing top applications across the industries are reviewed in detail. Moreover, the future and today's challenges of DNNs are also elaborated in this book.

Chapter 2

Deep Learning Basics

2.1 Applied Math

Nowadays, with access to many open-source ML libraries like TensorFlow, we can very easily create an NN without thinking about the underlying math. However, in order to fine-tune DL models or hyperparameters, along with right architecture selection, we need to fully understand how NNs and DL work. Therefore, in this section, we will talk about the basic mathematical concepts required to understand NN and DL in general. In order to understand the theory behind DL, the concepts of linear algebra are especially important [67].

The DL learning algorithm determines parameters, so-called weights and biases, using forward and backward propagation. The process of passing data through the NN layers is called forward propagation. Then, after each forward pass through NN, the backpropagation algorithm performs a backward pass to adjust the parameters of the model. To understand these basic processes in the learning phase, we will begin by describing the steps for both forward and backward propagation. We will then describe the methods currently used to optimize DL models.

2.1.1 Forward Propagation

For each neuron in the hidden or output layer, forward propagation occurs in two stages: preactivation and activation. The Fig. 2.1 shows a simple case with one hidden layer, however the explanation below applies to larger models.

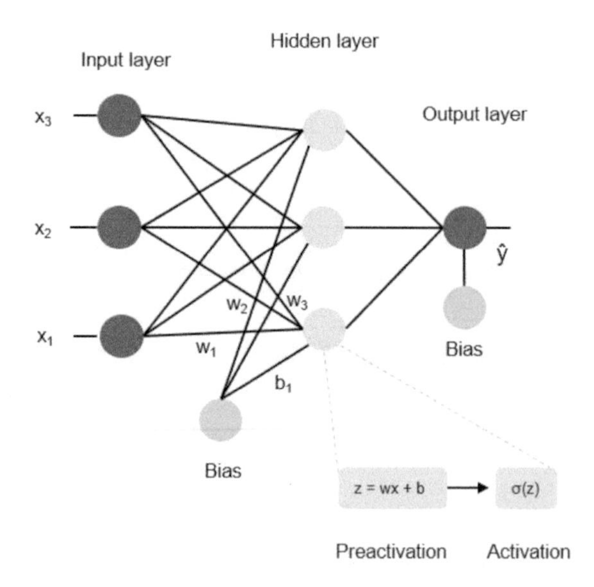

Figure 2.1: The forward propagation scheme.

Preactivation is a weighted sum of the input data. It represents the strength of the connection between neurons and the influence of the input on the final result. In simpler words, if the w_1 weight has a higher value than the w_2 weight, then the x_1 input will have a larger impact on the output. The preactivation phase also takes into account the bias known as an offset. Bias is required to move the activation function to the left/right to generate the required output values. For example, let the vectors of inputs and weights be $x = [x_1, x_2, ..., x_n]$ and $w = [w_1, w_2, ..., w_n]$ respectively. If the bias is b, then the preactivation function can be represented as:

$$z = x \cdot w + b$$

where $x \cdot w$ is the dot product of the inputs and the weights:

$$x \cdot w = (x_1 \times w_1) + (x_1 \times w_1) + ... + (x_n \times w_n)$$

After preactivation, the next step is the activation process based on the calculated weighted sum (Fig. 2.1). The activation function is required to add non-linearity to the outputs of the neurons, otherwise NN will become a linear transformation of the input data. This non-linearity provides great flexibility to NN through complex functions in the learning process. It also significantly affects the speed of the learning process, which is one of the main points for their selection. Several popular activation functions are in use today such as sigmoid,

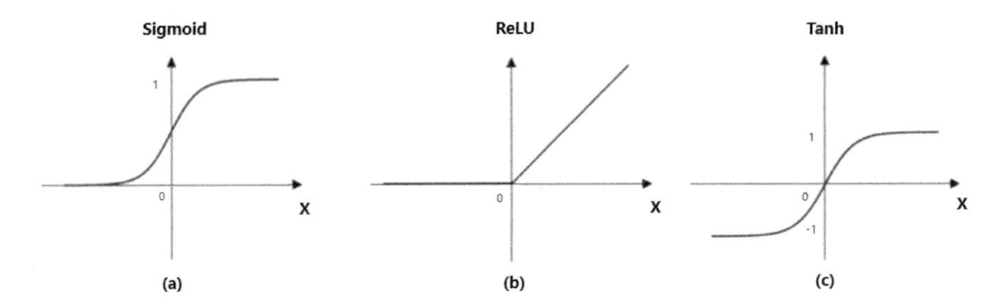

Figure 2.2: Most popular activation functions.

hyperbolic tangent (Tanh), rectified linear unit activation function (ReLU), etc. [182]. Their functional representations are shown in the Fig. 2.2—sigmoid (a), ReLU (b), Tanh (c).

Sigmoid and Tanh activation functions are mainly used for classification problems. If σ denotes a sigmoid or a Tanh function, then the output $\hat{y}$, known as the predicted value, will be defined accordingly as:

$$\hat{y} = \sigma(z_{sigmoid}) = \frac{1}{1+e^{-z}}, \sigma(z_{tanh}) = \frac{e^{z} - e^{-z}}{e^{z} + e^{-z}}$$

However, these linear activation functions are rather reluctant to allow gradient flow when they reach their saturation zones, which greatly reduces the learning potential of NNs. When multiple layers use certain activation functions in the NN, the gradients of the loss function approach to zero, making it difficult to train the network. This problem is called the vanishing gradient problem.

For example, the sigmoid activation function shrinks a large input space into a small input space between $[0, 1]$ (Fig. 2.3). Therefore, a large change in input

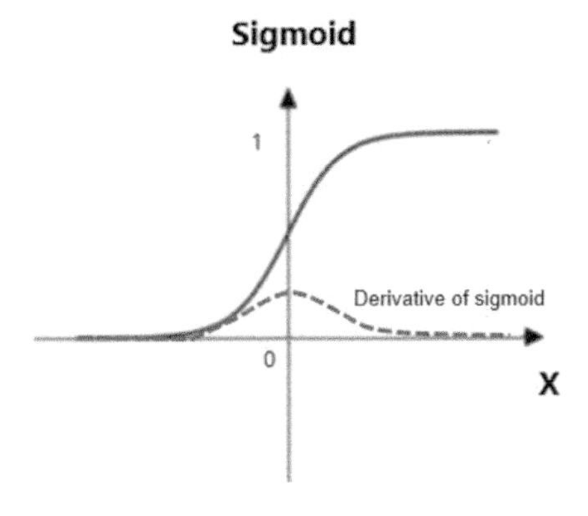

Figure 2.3: The sigmoid function with it's derivative.

causes a small change in output, and the derivative becomes small. In Fig. 2.3, when the input becomes larger or smaller, the derivative gets closer to 0. This problem is not important for shallow NNs, however, for DNNs, the small derivatives are multiplied together, which makes the gradient to be too small to train it effectively.

Fortunately, this problem has been resolved with the usage of ReLUs. Currently, for DL hidden layers, the most popular activation function is ReLU, which helps train much deeper NNs. In addition, from an optimization standpoint, ReLUs help solve the long standing problem of the vanishing gradient.

$$ReLU(z) = \begin{cases} z, & z > 0 \\ 0, & \text{otherwise} \end{cases}$$

2.1.2 Backward Propagation

DNN training is the most complicated part of DL due to the required large data set and large amount of computational power. A backward propagation algorithm, known as backpropagation, is used to effectively train NN through a technique called Chain Rule. The Chain Rule is designed to find the derivatives of the cost with respect to any variable [10]. Overall, the problem of learning is an optimization problem, and the backpropagation algorithm is used to navigate the space of possible sets of weights and biases that the model can use in order to make good predictions. To understand if we have achieved the desired predicted value, a loss function is used.

Currently, there are many types of loss functions to choose from when designing and configuring the NN model [168]. Generally, the choice of the loss function is directly related to the activation function used at the output of NN. Some of the currently popular loss functions are described below.

- The Mean Squared Error (MSE) is the most commonly used regression loss function, that represents the square of the difference between the actual (y) and predicted values ($\hat{y}$):

  $$L_{MSE} = (y - \hat{y})^2$$

  MSE values close to zero are better because in this case the model has less error. The disadvantage of this loss function is that outliers are not handled properly. Due to the squaring, the error will be quite large.

- Huber loss function (δ) is used in robust regression and is less sensitive to outliers. If the loss is too large, the quadratic equation changes directly to linear:

$$L_\delta = \begin{cases} \frac{1}{2}(y-\hat{y})^2, & |y-\hat{y}| \leq \delta, \\ \delta|y-\hat{y}| - \frac{1}{2}\delta^2, & otherwise. \end{cases}$$

The disadvantage of the Huber loss function is the additional training requirement to optimize the new δ hyperparameter in order to maximize the accuracy of NN model.

- Binary Cross Entropy (BCE) is used for binary classifiers. It shows the number of misclassified data points. The Cross Entropy (CE) loss function will be defined as:

$$L_{CE} = -\frac{1}{n}\sum_{i=1}^{n}(y_i log\hat{y}_i + (1-y_i)log(1-\hat{y}_i))$$

For binary classification, it would look like:

$$L_{BCE} = -\frac{1}{n}\sum_{i=1}^{n} log(1-\hat{y}_i),\ when\ y=0$$

$$L_{BCE} = -\frac{1}{n}\sum_{i=1}^{n} y_i log\hat{y}_i,\ when\ y=1$$

Here the sigmoid function will be used for calculating the output:

$$\hat{y} = \frac{1}{1+e^{-z}}$$

- Categorical Cross Entropy (CCE) is used for multiclass classification problems. If the number of categories is i and the number of rows is r, then the loss function for CCE will be defined as:

$$L_{CCE} = \sum_{i=1}^{c} y_{rj} log(\hat{y}_{rj})$$

For multiclass classification problems, the softmax function will be used for calculating the output:

$$\hat{y}_{rj} = \frac{e^{z_r}}{\sum_{j=1}^{K} e^{z_j}}$$

where it takes as input the vector z of K real numbers.

While the loss function is defined as an error at one data point, the cost error function is the sum of all errors in the entire data set. Therefore, the cost function in the case of MSE will be defined as:

$$C_{MSE} = \frac{1}{n}\sum_{i=1}^{n}(y-\hat{y})^2 \rightarrow min$$

The goal of backpropagation is to minimize the cost function with respect to the weights and bias. To find the optimal weights and bias for the NNs, the gradient of the cost function will be used. The gradient of the function C at the point a is the vector of partial derivatives C at $a = [a_1, a_2, ..., a_n]$, which shows the rate of the cost function change with respect to the change in its argument (a).

$$\frac{\partial C}{\partial a} = [\frac{\partial C}{\partial a_1}, \frac{\partial C}{\partial a_2}, ..., \frac{\partial C}{\partial a_n}]$$

In simple words, the gradient shows how much the input parameters (weights and bias) need to be changed in order to minimize the loss function C. Therefore, the computation of gradients will be done using the Chain Rule [10].

$$\frac{\partial C}{\partial w} = \frac{\partial C}{\partial \hat{y}} \times \frac{\partial \hat{y}}{\partial z} \times \frac{\partial z}{\partial w}$$

where the gradient of the various components is defined as:

$$\frac{\partial C}{\partial \hat{y}} = \frac{2}{n} \times \sum (y-\hat{y})$$

$$\frac{\partial \hat{y}}{\partial z} = \sigma(z) \times (1-\sigma(z))$$

$$\frac{\partial z}{\partial w} = \frac{\partial}{\partial w_i}\sum_{i=1}^{n}(x_i \cdot w_i + b) = x_i$$

From the equations above, the gradient of the cost function C with respect to the weights w can be defined as:

$$\frac{\partial C}{\partial w} = \frac{2}{n} \times \sum (y - \hat{y}) \times \sigma(z) \times (1 - \sigma(z)) \times x$$

Since the bias is theoretically considered to have an input of constant value 1, the gradient of the cost function C with respect to the bias b can be defined as:

$$\frac{\partial C}{\partial b} = \frac{2}{n} \times \sum (y - \hat{y}) \times \sigma(z) \times (1 - \sigma(z))$$

To optimize the selection of weights and bias, we can use a gradient descent that changes these parameters proportional to the negative of the gradient of the cost function with respect to the corresponding weight or bias. So, the algorithm of gradient descent flow can be described in the following steps:

1. Calculate the gradient at the current data point x_1.

2. Take a small step, called learning rate (α), in the direction of the gradient to reach the point x_2. In other words, α is a parameter that is used to control how much the weights and bias are changed. Hence, the weights and bias are updated as shown below:

$$w = w - (\alpha \times \frac{\partial C}{\partial w})$$

$$b = b - (\alpha \times \frac{\partial C}{\partial b})$$

3. Repeat the entire process until convergence.

There are several processes in this algorithm that have improved over the years. Modern gradient descent based optimizers use a variety of techniques such as adaptive step size, momentum, and so on, which we will discuss shortly in the next section.

2.1.3 Optimization of DNNs

As we discussed earlier, DNNs are efficient learning approaches that are widely used in DL. They automatically extract features from large unstructured data, and then learn the mapping between these features in a hierarchical manner. This very deep hierarchy enables DNNs to perform with high accuracy in many learning tasks.

Building on the knowledge gained from previous discussions, in this section, we will try to summarize the main parts of DNNs that require a lot of attention when building a model. While much research has been done towards the automatic detection of the correct NN components there is still a lot of work to be done. Below are a few important aspects to consider when building and training DNNs.

Normalized initialization

When training DNN, it is very important to adequately initialize the weights and biases for each layer. The goal of weight initialization is to prevent the layer activation outputs from vanishing during the forward pass through a DNN layer. If vanishing occurs, the loss gradients will either be too large or too small to flow backwards beneficially, and the network will never converge. Consequently, if the weights are initialized too large or too small, the network will not learn well. Hence, it is very important for DNN to initialize the weights correctly.

The challenge of normalized initialization was first discussed in the work [62] and by the name of the author it is called Xavier's initialization. In the work, the authors derived an appropriate range of uniform distribution for each layer, which is used to sample the weights for the network initialization. They stated that the normalization factor is very important when initializing DNNs due to the multiplicative effect through layers. For maintaining activation variances and backpropagated gradients variance as one moves up or down the network, they set the weights to values chosen from a random uniform distribution that is limited between:

$$\left[-\frac{\sqrt{6}}{\sqrt{n_i+n_{i+1}}}, \frac{\sqrt{6}}{\sqrt{n_i+n_{i+1}}}\right]$$

Here, n_i is the incoming network connections number and n_{i+1} is the outgoing network connections number from that layer. The authors have proven that Xavier's initialization results in quicker convergence and higher accuracy.

However, this conclusion is based on the assumption that the activation functions are linear. Consequently, this approach leads to poor results when working with non-linear activations such as ReLU. For ReLU activation, a single layer will have a standard deviation that is very close to the square root of the input connections number, divided by the square root of two. Scaling the values of the weight matrix will result each individual layer having a standard deviation of 1 on average. Therefore, in the work [77], the authors investigate the possibility of the best initialization of weights in DNNs with ReLU kind of asymmetric activation functions. The authors proved that DNNs converge much earlier with the

Kaiming initialization, where each randomly chosen number from the standard normal distribution is multiplied by $\sqrt{\frac{2}{n}}$, where n is the incoming connections number.

In general practice, biases are initialized with zeros, since the asymmetry breaking is provided by the small random numbers in the weights. For ReLU non-linearities, some researchers use small constant values such as 0.001 for all biases because this ensures that all ReLU units fire in the beginning. Hence, selecting the right combination of weight initialization method and bias initialization can be critical in training your DNN model.

Network regularization

The main goal of ML and DL is to have a model that performs well not only with the training data set, but also with the new data that will be used for predictions. Performing well with unobserved inputs is called generalization. Network regularization is a set of techniques used by DNNs to keep the model generalization and improve the accuracy of the model.

Figure 2.4 shows a nice example of model fitting. The underfitting of the model refers to the case when the model fails to sufficiently learn the problem and does not perform well even on the training data set. For overfitting, the model learns the training data set very well, hence performing well, however it does not work on the new data when predicting the output. The appropriate fitting learns well on the training data set and generalizes well on the new unseen data set as well. With regard to bias and variance, we can say that the underfitting model has high bias and low variance, while the overfitting model has low bias and high variance. The underfitting problem can be solved by increasing the capacity of the model in the way that the model can fit more types of functions for mapping inputs to outputs. This can be achieved by changing the structure of the model in

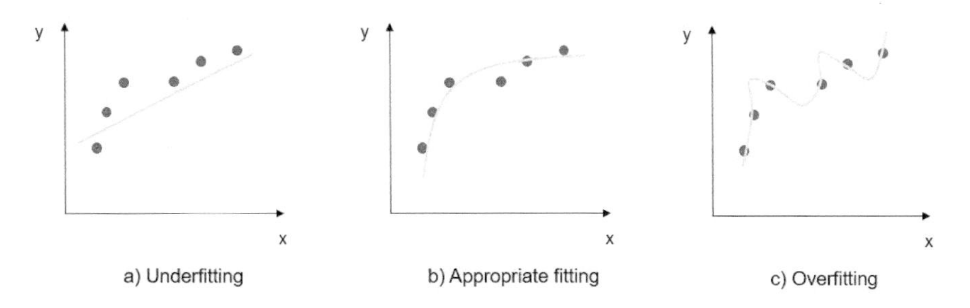

Figure 2.4: Model fitting example.

terms of adding additional hidden layers or changing the number of nodes within the layers.

The problem of overfitting is much more complicated compared to underfitting. The easiest way to reduce overfitting is to increase the size of the training data. However, since the labeled data is too costly, it is quite hard to increase the size of the training data. To solve this issue, regularization of the network is used. In DL, regularization actually penalizes the weight matrices of the nodes. The coefficient of regularization value needs to be optimized in order to obtain an appropriate fitting model.

There are different techniques for this purpose [85]. And the common regularization techniques are $L1$ and $L2$, which update the cost function by adding a regularization term. Consequently, the values of the weight matrices decrease, which to some extent reduces the overfitting of the model.

For $L1$ and $L2$ (also known as weight decay) regularization, the cost function will be defined as follows:

$$C_{L1} = F_{Loss} + \frac{\lambda}{2n} \sum ||w||$$

$$C_{L2} = F_{Loss} + \frac{\lambda}{2n} \sum ||w||^2$$

where λ is the regularization hyperparameter that needs to be optimized for best results. Unfortunately, the $L1$ and $L2$ regularization may restrict the NNs to utilize their full potential.

Another common type of regularization that is widely used in DL is dropout [186]. The key idea of dropout is selecting and removing random nodes at every iteration. This results in a different set of outputs. The probability of choosing the number of nodes that should be dropped is a hyperparameter of the dropout function that also needs to be optimized for best results.

To understand when a model needs regularization, the plotting learning curves need to be examined. Graphing out the model's training and validation errors is the most widely used method for determining whether a model has overfitted or not (Fig. 2.5).

If the model cannot generalize well on previously unobserved data, then the corresponding loss/error value will inevitably be high. Therefore, in the case of an overfitting model, the training error and validation error will be very different, moreover, the validation error will always be higher than the training error. There is a cross-validation strategy called early stopping, in which one part of the training set is stored as a validation set. When the performance on the validation set deteriorates slightly, training should be stopped, since after that the model will start overfitting on the training data.

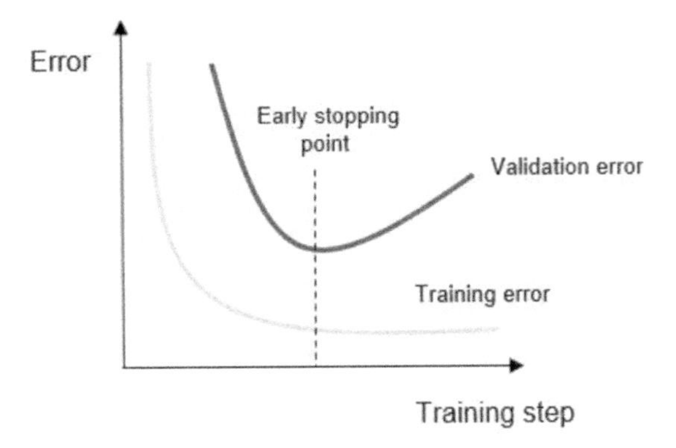

Figure 2.5: Learning curve of overfitting model.

Data scaling

These are other effective data preprocessing techniques for optimizing deep networks. DNNs use gradient descent where the input feature can affect the step size of the gradient descent. Consequently, differences in the feature ranges may lead to different step sizes for each feature. To ensure that these steps are updated at the same rate, the data need to be scaled before training the model. For this purpose, data normalization is designed to rescale the input and output variables prior to training NNs, which can significantly improve stability and performance.

Differences in the input data scales may increase the difficulty of model training. For example, large input values can lead to a model that uses large weights, which can result in an unstable model with a higher generalization error. There are two types of data scaling: normalization and standardization. Normalization is the rescaling of the input data from the original range within the range $[0,1]$. This requires to have accurately estimated minimum and maximum values in the data, since the value x will be normalized as follows:

$$y = \frac{x - min}{max - min}$$

Data standardization involves rescaling the distribution of values so that the mean of observed values is 0 while the standard deviation is 1. This can be seen as centering of the data. Therefore, data standardization requires to have accurately estimated mean and standard deviation of the values, since the value x would be standardized as follows:

$$y = \frac{(x - mean)}{standard\_deviation}$$

There is no right answer to the question of when to use normalization over standardization and vice versa. It mainly depends on the data and algorithm used in DNNs. In general, normalization is useful when the distribution of the input data does not follow the Gaussian distribution. On the other hand, standardization can also be helpful in cases where the data follows a Gaussian distribution. In addition, even if there are outliers in your data, they will not be affected by standardization as opposed to normalization.

The idea of normalizing inputs for each layer of DNNs was first presented in the work [90]. The authors suggested Batch Normalization that addresses the internal covariate shift by normalizing the DNN layer inputs. It scales the output of the layer by standardizing the activations of each input per mini-batch. This stabilizes and speeds up the training process of DNNs. The authors stated that the normalization step should be applied immediately before the nonlinear function. However, some researchers stated also that it could be applied afterwards. On each hidden layer, the Batch Normalization flow first determines the mean μ and the standard deviation σ based on the activation vector Z:

$$\mu = \frac{1}{n}\sum_i Z^i$$

$$\sigma = \frac{1}{n}\sum_i (Z^i - \mu)$$

It then normalizes the activation vector in the way, that each output follows a standard normal distribution across the batch:

$$Z^i_{norm} = \frac{Z^i - \mu}{\sqrt{\sigma^2 - \varepsilon}}$$

Finally, it applies two parameters γ and β to calculate the Z output:

$$Z = \gamma * Z^i_{norm} + \beta$$

These two parameters γ and β allow to adjust the standard deviation and the bias. Thus, during the calculations, these parameters need to be trained through gradient descent method. Since Batch Normalization sometimes reduces generalization error and allows dropout to be omitted, it is not recommended to use it with dropout for the same DNN [64].

Residual learning

The depth of network plays an important role in NN architectures as more layers progressively learn more complex features. Deeper networks, however, are very difficult to train. In their work [78], the authors presented that the training error of a 56-layer network is higher than that of a 20-layer CNN. This failure of deep networks could be explained by an optimization function or a vanishing gradient problem. However, in the same work, the authors stated that the use of Batch Normalization ensures that the gradients have healthy norms. They suggested the residual learning that eases the training of DNNs, resulting in better performance. These residual networks are much deeper, however, they require a similar amount of weights [78].

These residual networks provide residual connections directly to the earlier layers (Fig. 2.6), which allows for the input x and $F(x)$ to be combined as input to the next layer. These connections lead to better information flow during the forward propagation and an effective gradient flow during the backpropagation. From an optimization standpoint, residual learning simply eliminates the vanishing gradients problem and makes it theoretically possible to train very deep networks.

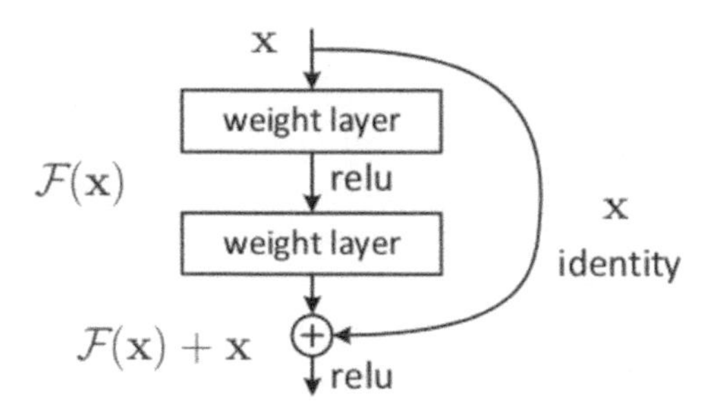

Figure 2.6: A residual block—https://arxiv.org/pdf/1512.03385.pdf.

Network configuration

To configure NNs, two main hyperparameters that control the network architecture need to be defined. Those are the number of layers and the number of nodes for each layer. As discussed earlier, residual networks make it possible to train even very deep networks. However, the question still remains: "How do we need to define the number of layers and the number of nodes?". We we will try to address this question in this part.

Several terms are often used when talking about the architecture of NNs:

- Width - the number of nodes in layers,
- Size - the number of nodes in overall model,
- Depth - the number of layers in the NN,
- Capacity - the type or structure of functions that can be learned,
- Architecture - the arrangement of the layers and nodes in the NN.

In our discussions, when we are talking about the depth of the NN, the input layer will not be counted as it is simply associated with the input variables. Hence, if the network consists of an input layer, one hidden layer and an output layer, this means we have a two-layer NN. In general, the width of the input layer is equal to the number of features in the input data. Some NN configurations add one more additional node for bias. A single layer NN can only be used to represent linearly separable functions. In other words, for relatively simple problems a single layer network would be sufficient.

For complex problems, it is necessary to design multi-layer NNs. In [126] work, it is stated that two hidden layers are more than enough for creating classification regions of any shape. On the other hand, in [64], the authors insisted that with one hidden layer a feedforward network can approximate any function. However, even for those networks with sufficiently large one hidden layer, it may be more efficient to examine it with a large number of hidden layers.

There are several theoretical investigations in the direction of finding the optimal depth and width for DNNs [163, 79, 181]. However there are no yet clear conclusions on how to define neural architecture for maximum performance. Similar to what we have previously discussed for the width of the input layer, the width of the output layer is also completely determined by the configuration of the chosen model (classifier, regression).

Hence, the main problem is related to determining the width of the hidden layers. In [79], the author suggested several ways to select the width of hidden layers. Deciding and configuring the right number of neurons is quite critical. Using too few neurons in hidden layers will result in underfitting, preventing signals from being detected in a complex data set. On the other hand, too many neurons in hidden layers can lead to overfitting. In addition, a large number of neurons in hidden layers can increase the time required to train the network. Therefore, there should be some average value for the width of the hidden layer so as not to create additional problems.

One of the common things is that the optimal width of the hidden layer is usually between the width of the input and output layers. Each hidden neuron added to the NN will increase the number of weights that will add more complexity to the system. Hence, it is recommended to use the least number of hidden neurons that accomplish the task. There are several methods for determining an acceptable number of neurons in hidden layers, such as the following:

- The number of hidden neurons should be between the size of the input layer and the size of the output layer,

$$N_i < N_h < N_o$$

- The number of hidden neurons should be less than twice the size of the input layer,

$$N_h < 2N_i$$

- The number of hidden neurons can be defined based on the number of samples in the training set,

$$N_h = \frac{N_s}{\alpha * (N_i + N_o)}$$

  where N_i and N_o are the number of neurons in the input and output layers, respectively. N_s is the number of samples in the training data set. α is a hyperparameter that needs to be tuned, usually in the range $[2, 10]$.

However, these rules are just a starting point. In general, there is no right way to analytically calculate the depth and width of a network to address a specific predictive modeling problem in the real world. Right configuration is the result of systematic experimentation with the input data set.

Optimizers

In this part, we will discuss various algorithms that change the attributes of NNs to reduce losses. In general, these methods change the weights and learning rate of the NNs and are responsible not only for reducing losses, but also for providing the most accurate results.

We have already reviewed the Gradient Descent algorithm, which is widely used in regression, classification problems and backpropagation of NNs [171]. It is an approximate and iterative method for mathematical optimization that is used in NNs for finding weights and biases. Many improvements have been made to the Gradient Descent algorithm over the years. One of the variants is Stochastic

Gradient Descent, which updates the parameters of the model more frequently, and therefore, converges in less time. For example, if for a chosen point of vector $\theta = (\theta_1, ..., \theta_n)$, the Gradient Descent algorithm for the θ parameter is defined as:

$$\theta = \theta - \alpha \nabla C(\theta)$$

Then the Stochastic Gradient Descent algorithm will look like:

$$\theta = \theta - \alpha \nabla C(\theta; x_i; y_i)$$

where α is a small positive value called learning rate, x_i and y_i are the training examples. However, the Stochastic Gradient Descent algorithm has a high variance in the model parameters. To solve the variance challenge, a Mini-Batch Gradient Descent algorithm can be used, in which the parameters are updated after each batch into which the data set is divided [107]:

$$\theta = \theta - \alpha \nabla C(\theta; B(i))$$

where $B(i)$ are the batches of training data sets.

However, all Gradient Descent algorithms have the following challenges:

- Avoiding getting trapped in local minima that can be a result of saddle points (where one dimension slopes up and the other slopes down) [41],
- Choosing a proper value of the learning rate α,
  - If α is too small, it may take a long time for gradient descent to converge,
  - If α is too large, it may fail to converge and overshoot the minimum.
- Choosing the same learning rate for all parameters.

To overcome these challenges, there are several algorithms that are widely used in the DL community nowadays. Momentum is an algorithm that is mainly used to reduce the high variance in Stochastic Gradient Descent, thereby softening the convergence [156]. For this algorithm, a new hyperparameter, that is called momentum γ, is used to add a change into the current update vector from the previous one:

$$\theta = \theta - V(t)$$

$$V(t) = \gamma V(t-1) + \alpha \nabla C(\theta)$$

The disadvantage of the momentum method is in defining the new hyperparameter accurately. It is usually set around 0.9 value, smaller than 1. The momentum is increased for dimensions in which gradients point in the same directions, and vice versa, resulting in faster convergence and less oscillation. However, if the momentum is too high, the algorithm may miss the local minima and continue to rise. This issue has been addressed in Nesterov Accelerated Gradient algorithm [141]. It used the idea of calculating the gradient not at the current position but at the approximate future position of the parameters, which can be given as $\theta - \gamma v_{t-1}$:

$$\theta = \theta - V(t)$$

$$V(t) = \gamma V(t-1) + \alpha \nabla C(\theta - \gamma V(t-1))$$

One of the important challenges of choosing the same learning rate for all parameters has been addressed in the Adagrad algorithm [46], which was later extended to Adadelta [226]. These methods change the learning rate for each parameter at a given time step. Updates are different based on the occurrence of the parameter features. For parameters with a large number of feature occurrences, they perform smaller updates and vice versa. Hence, these methods give better results for sparse data while other methods perform poorly. If g_t is the gradient at the t time step and $g_{t,i}$ is the partial derivative, then the update for each parameter at the time step for Adagrad will be:

$$\theta_{t+1,i} = \theta_{t,i} - \frac{\alpha}{\sqrt{G_{t,ii} + \varepsilon}} g_{t,i}$$

$$g_{t,i} = \nabla_\theta C(\theta_{t,i})$$

where G_t is a diagonal matrix with the sum of squares of gradients, and ε is a term for avoiding division by zero.

Adadelta algorithm limits the window of accumulated past gradients to some fixed size in order to reduce the learning rate:

$$\theta_{t+1} = \theta_t - \frac{\alpha}{\sqrt{E[g^2]_t + \varepsilon}} g_t$$

$$E[g^2]_t = \gamma E[g^2]_{t-1} + (1-\gamma) g_t^2$$

Therefore, the running average $E[g^2]_t$ depends only on the previous average and the current gradient.

One of the most common used optimizers that calculates the learning rates for each parameter is Adaptive Moment Estimation (Adam) [110]. This method also preserves the exponentially decaying average of past gradients m_t. The following formula is used to update the parameters:

$$\theta_{t+1} = \theta_t - \frac{\alpha}{\sqrt{\hat{v}_t + \varepsilon}} \hat{m}_t$$

where m_t and v_t are values of the mean and non-centered variance of the gradients, respectively:

$$\hat{m}_t = \frac{m_t}{1 - \beta_1^t}$$

$$\hat{v}_t = \frac{v_t}{1 - \beta_2^t}$$

In the work, the authors suggest default values such as 0.9 for β_1, 0.999 for β_2 and 10^{-8} for ε. In addition, they show that this algorithm overcomes other similar approaches.

Choosing the right optimizer for NN actually depends on the data set and model. For sparse input data, the best results will be obtained using one of the adaptive learning rate methods. In addition, usually the default values of learning rates suggested for each optimizer are likely to give the best result, so there is no need to tune them separately.

Chapter 3

Neural Network Structures

Different types of neural network (NN) architectures exist, though they are generally based on NNs with multiple layers of stacked neurons that allows the back-propagation of a signal. Deep models that employ feed-forward layers are referred to as a multi-layered perceptron or MLP.

NNs can be viewed broadly as generative or discriminative. A discriminative NN model discriminates between different kinds of data instances, for example, it could differentiate a dog from a cat. Generative models on the other hand can generate new data instances, for example, generating new photos of animals that look like real animals. Generative Adversarial Networks are a type of generative model. In essence, given a set of data instances X and a set of labels Y, generative models capture the joint probability $p(X,Y)$, or just $p(X)$ if there are no labels whereas discriminative models capture the conditional probability p(Y — X). Predictive models (for example those that predict the next word in a sequence) are typically generative models; they assign a probability to a sequence of words. Discriminative model on the other hand do not address likelihood of an instance, but rather that of a label to apply to the instance.

Generative models can capture correlations rather than just differences. As such, they can tackle more complicated tasks than comparable discriminative models. For example, a generative model can capture a distribution like "noses are unlikely to appear above the eyes." or "things that look like cars are probably going to also have things that look like steering wheels". A discriminative model, by contrast, will be able to tell the difference between "face" or "not face" by examining patterns while overlooking the complicated correlations that are important in a generative model [199]. Within the data space, generative models show how data is placed throughout, while discriminative models attempt to create boundaries within the space. Classification is traditionally referred to as discriminative modelling; that is, using the training data to find a discriminant function $f(x)$ that maps each x directly onto a class label, such that the inference and decision stages are combined into a single learning problem. Well-known examples of generative NN include the Generative Adversarial Networks (GAN) and Deep Belief Networks (DBN).

In terms of network learning mode, two of the commonest deep learning architectures [127] include:

- systems based on supervised end-to-end training of an entire network. Recurrent neural networks (RNNs) and convolutional neural networks (CNNs) fall in this category, with CNNs being the most well-known architecture within image processing in recent times. AlexNet [16, 118, 224] is the most well-known, general classification CNN architecture [127].

- systems based on unsupervised, layer-by-layer pre-training of networks, with supervised fine-tuning of the network. Deep belief networks (DBNs), stacked auto-encoders (SAEs) and restricted Boltzmann machines (RBMs) are in this category.

Most pre-trained NN models in deep learning can be broadly grouped into Artificial Neural Networks (ANN), Convolutional Neural Networks (CNN) and Recurrent Neural Networks (RNN). However, there are also NNs that do not fall directly into any of these classes because they operate using an integration of multiple network types. Examples of such types of NN models are generative models like generative adversarial networks (GAN) and Deep Belief Network (DBN). As NN types continue to grow exponentially, keeping up with the various types of architectures that are continually emerging gets more difficult. Figure 3.1 provides a neat summary of NN typologies by Veen [201]. Some of these architectures will be discussed in more details in the following sections.

3.1 Feed Forward Neural Network

This is one of the simplest forms of NN; a front propagated wave with no back-propagation. The data (input) travels in only one direction; it passes through the

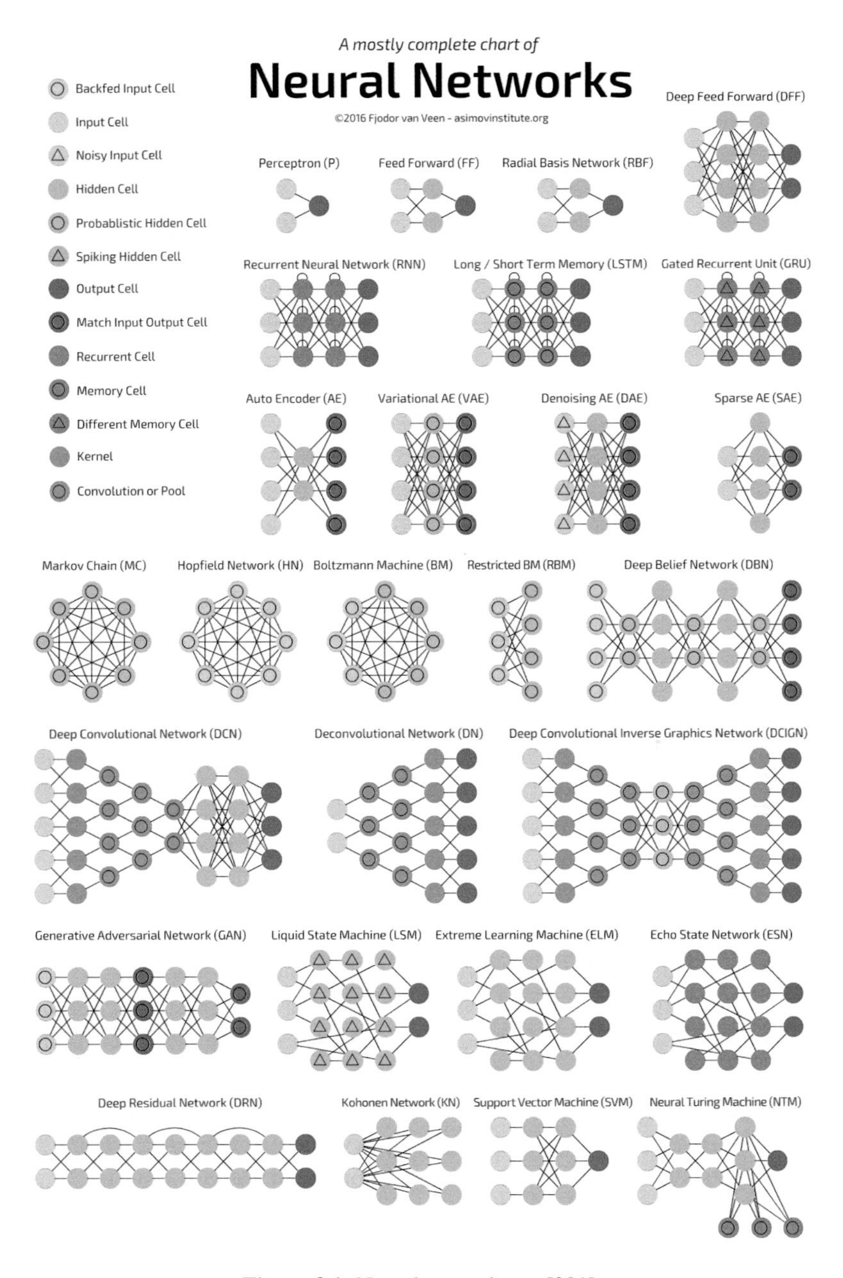

Figure 3.1: Neural network zoo [201].

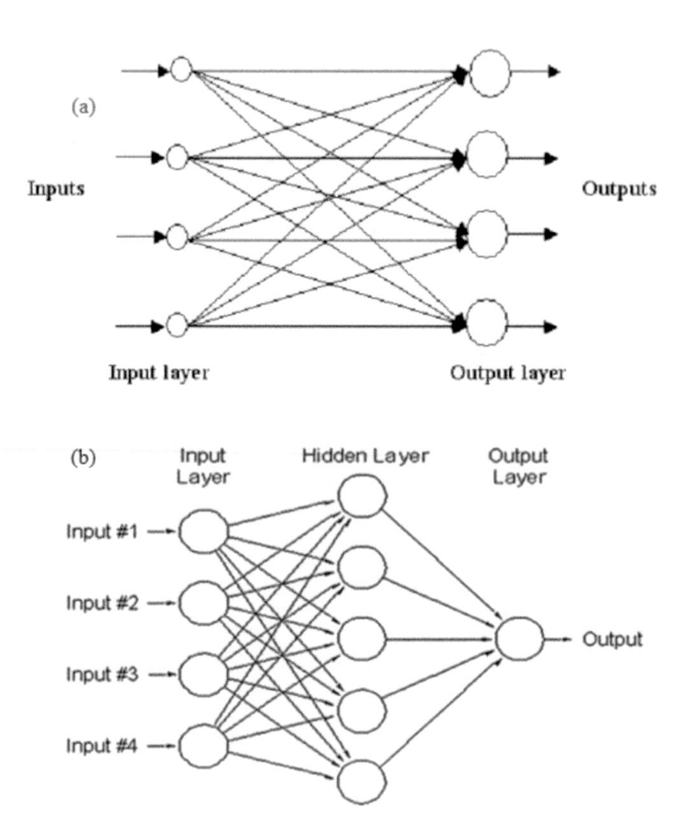

Figure 3.2: Single layer and multi-layer feedforward network [176].

input nodes and exits at the output nodes. In feed-forward NN, all nodes are fully connected, activation flow is one-directional, from input to output layer and no back loops and only one hidden layer between them. Training usually employs back-propagation. Figure 3.2 shows a single layer (top) and multi-layer (bottom) feed-forward NN. These types of networks are responsive to noisy data and they are easy to maintain. They are useful in situations where classifying the target classes is complicated; for example, in computer vision and speech recognition.

Multi-layer FF networks were almost impractical initially due to the setback of exponential growth of training times resulting from error passed on from previous layers. As new approaches for effective training of deep FF networks emerged, they have become central to modern machine learning systems, and in addition to being able to serve most of the purposes as FFs, they provide much better results.

3.2 Convolutional Neural Networks (CNN)

These are artificial models of mammalian visual cortex [121, 125] that has been validated by strong biological plausible evidence from the early works of Hubel and Wiesel [88] on the cat's visual cortex. They were inspired by the connectivity pattern of neurons in a mammal's nervous system. Within the visual cortex, each neuron responds to stimuli in the receptive field, a restricted region of the visual field. Multiple instances of such fields overlap to cover the entire visual area.

The domain of computer vision hopes to enable machines with capabilities to perceive and relate to the world the way humans do. Related tasks cover image recognition, analysis and classification, natural language processing, and recommender systems among others. CNN represents advancements in computer vision with DL that has been constructed and perfected with time. CNNs feature convolution cells (or pooling layers) and kernels. Convolution kernels are responsible for processing input data, while pooling layers simplify the input by using functions to reduce unnecessary features. CNNs are typically used for image recognition; the input window slides along the image, pixel by pixel while the data is passed to convolution layers which compresses detected features. CNNs have the ability to learn both local and global structures in images, and have shown outstanding usefulness in real world applications and in big data tasks related to pattern recognition. Due to their great effectiveness, CNNs have become state-of-the-art within the field of computer vision and are among the commonest DL architectures. CNNs have been found to outperform human experts in some domains [179].

CNNs are made of four layers: the sub-sampling layer (max-pooling), Rectified Linear Unit Layer (ReLU), the spatial convolutional layer, and a fully connected layer. CNN applies equal weights in the convolutional layers (equal weighted filter applied to each pixel of feature maps) and therefore requires less memory and attains higher processing speed [173]. In deep CNN, a layer may detect gradients, another detect shapes, another, lines, etc. to the scale of particular objects. Figure 3.3 presents the basic structure of a CNN, showing the four layers.

CNNs learn the features or filters which are then applied for object classification, and do not depend on prior knowledge and are also less dependent on hand-engineered features. This feature represents one of the main benefits of CNNs. Figure 3.4 is an example of a CNN sequence to classify handwritten digits.

In CNN, an input volume is normally convolved with several different kernels as each kernel extracts a different feature from the input image. In a single-filter convolution, a kernel is passed over an input volume (the image), and the

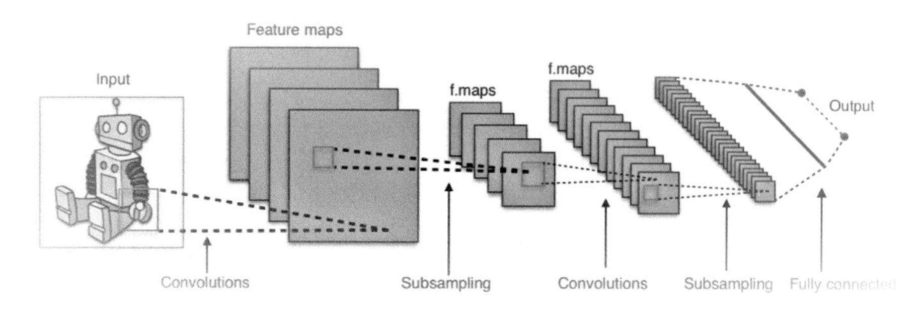

Figure 3.3: Convolutional neural network structure [174].

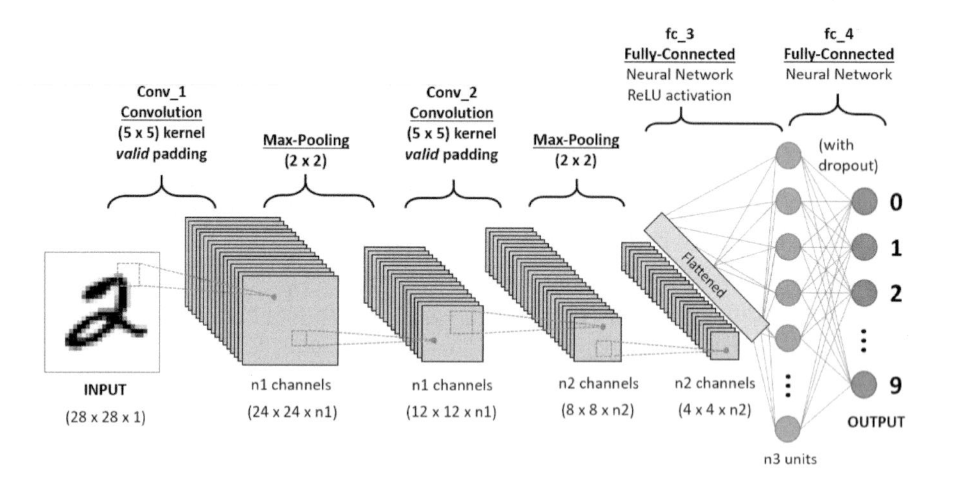

Figure 3.4: Example of a CNN sequence to classify handwritten digits [174].

values of the image that matches the size of the kernel, performs an element-wise matrix multiplication between them to generate the value of one cell on the output volume. The sum of progressive multiplication results goes into the feature map. Figure 3.5 shows an input and a kernel/filter unit and the resulting feature map generated from different convolutions (top and bottom).

The lighter grey area where the convolution operation takes place is the receptive field; it is equal in size $(3X3)$ to the filter/kernel. The kernel slides in the right direction until it reaches the last position. This sliding or stride, is 1 (single step) by default. Due to the nature of deep networks, If volume is decreased at every convolutional layer, that is, if larger strides are used to achieve overlap in the receptive region, the resulting feature map generated is smaller than the filter. In such cases, the resulting volumes may become too small to capture the intended features. To prevent this, padding is implemented. This is effectively

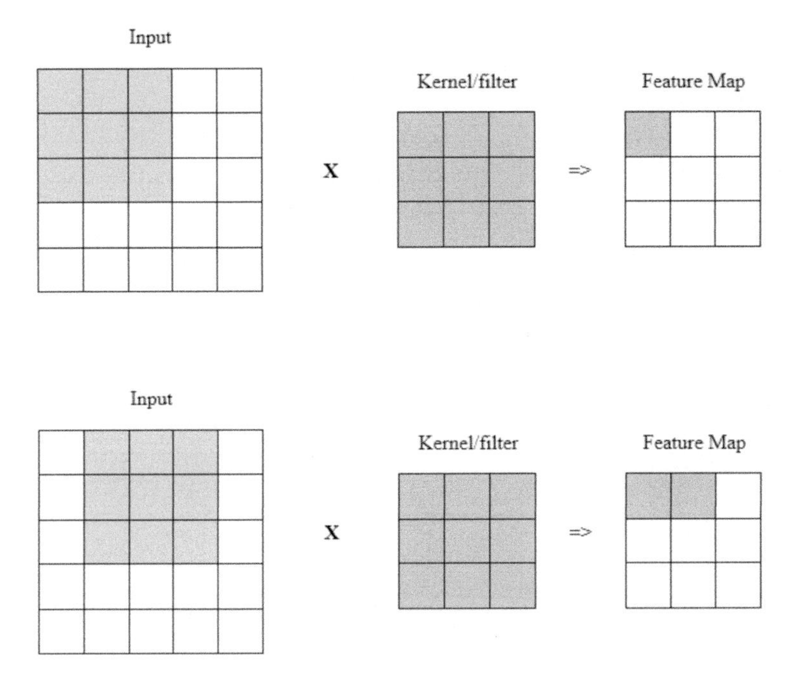

Figure 3.5: Input and a kernel/filter unit with resulting feature map for different convolutions.

a technique whereby zeros are added to the margin of the image to increase its size. Padding ensures the achievement of the same volume on both side of the convolution as shown in Fig. 3.6 (upper and lower). The stride remains one, but with the padding, the input gets to be 7 instead of 5, leading to an output size of 5 as shown.

The behavior of the output size is described by:

$$O = \frac{(I + 2P - K)}{S + 1}$$

Figure 3.7 provides a 3D view of the convolutions described above showing how the dimensions of the kernels are matched with that of the input volume, leading to a 1D plane output volume (a) or a 3D convolution with smaller kernels (b).

The number of filters determines the channel size dimension in the implementation of convolutions in the code. Hence, when convolutions are done with multiple filters/kernels, each one stacks a different output volume, with different features (edges, corners, depth, etc.) activated to capture each of these features

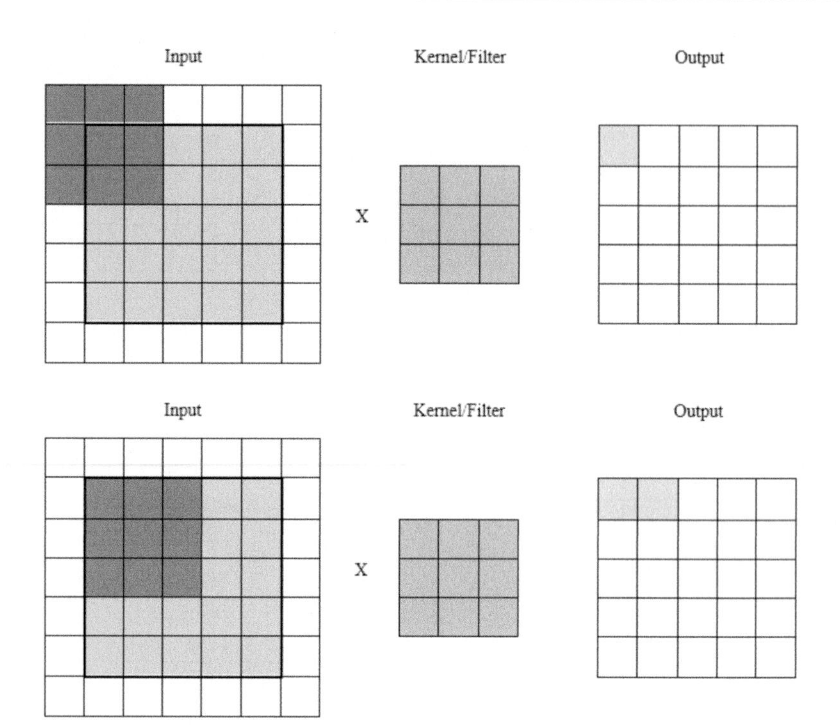

Figure 3.6: Convolution with 1 kernel, stride 1, padding 1.

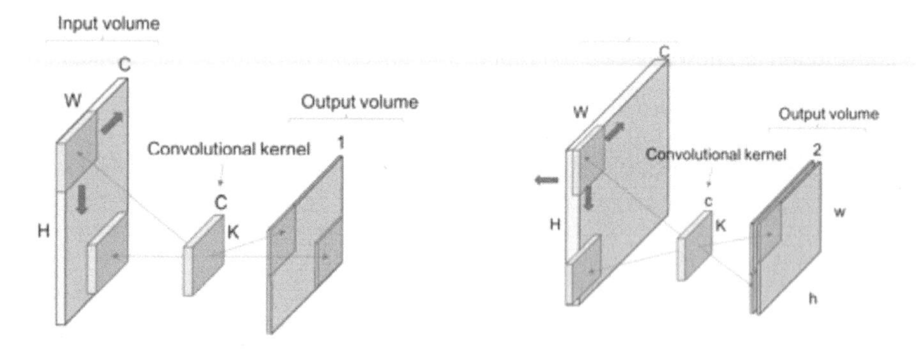

Figure 3.7: Convolutions with single slides (a) and smaller kernels (b).

as shown in Fig. 3.8. Each kernel is leading to one feature map in the output volume.

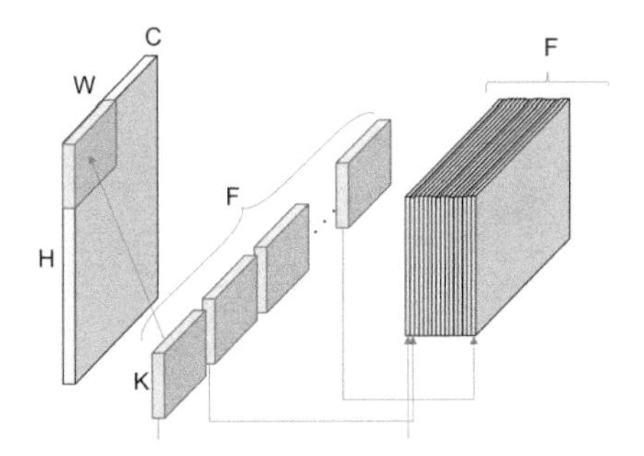

Figure 3.8: Convolution with multiple kernels.

3.3 Deconvolutional Neural Network (DNN)

De-convolutional Neural Network (DNN) is essentially reversed CNN. It is reading the CNN architecture from right to left, rather than left to right. Simply put, it is CNN that takes an image, classifies it and runs it in reverse, converting a class vector to an image. The DNN produces new images from scratch, the network does not learn the specific images, but pulls on perspectives from similar objects to produce a new image from scratch. It uses its previous knowledge to rotate the images correctly. DNN has applications in different kinds of objects from human and animal faces to landscapes, houses, cars and other types of objects. In other words, with DNN, any type of image can be generated. DNN is able to generate new images from photos of objects just by changing the angle of view; meaning for example that faces of non-existent people can be generated or novel facial expression, dressing, and poses can be generated from a person's image. DNN appears to be the basis of an unprecedented degree of image processing or photo-editing. Figure 3.9 shows the structure of a DNN. In its operation, it takes cat image, and produces vector like dog: 0, lizard: 0, horse: 0, cat: 1; it can then take this vector and draw a cat image from it.

Figure 3.10 shows the structure of an architecture, essentially a CNN. When the architecture is read from right to left, it takes an image of a chair and classifies it and when run in reverse (left to right), it will convert a class vector to a complete chair image.

3.4 Deep Convolutional Inverse Graphics Network (DCIGN)

The basic Deep Convolutional Inverse Graphics Network (DC-IGN) architecture looks essentially like a head-to-head unit of a CNN and a DNN. It however ac-

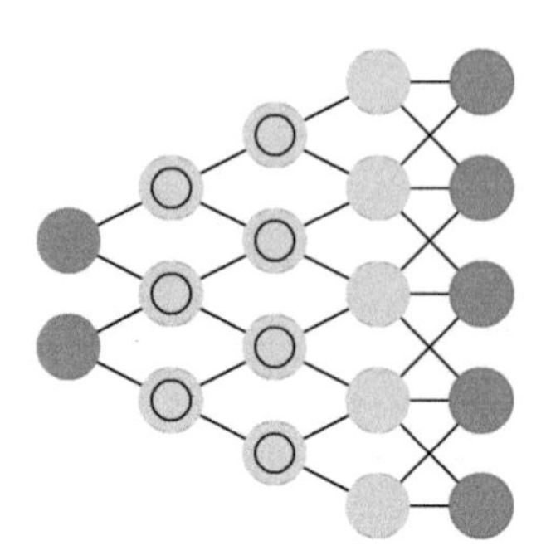

Figure 3.9: Structure of a DNN.

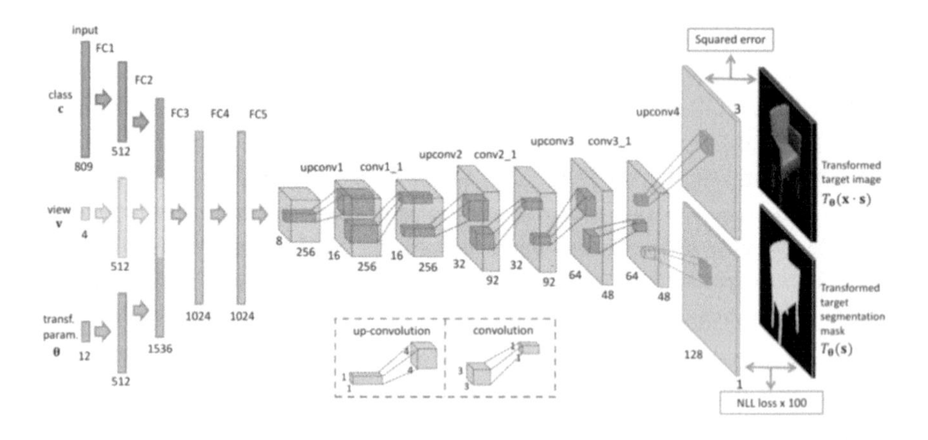

Figure 3.10: De-convolutional neural network structure [54].

tually consists of an encoder network and a decoder network (Fig. 3.11). DCN and DNN do not act as separate networks, instead, they are spacers for input and output of the network.

The encoder network captures distribution over graphics codes Z, given data x, while the decoder network learns a conditional distribution to produce an approximation x, given Z, which can be a disentangled representation containing a factored set of latent variables $ziEZ$ such as pose, light and shape. Due to this level of abstraction, DC-IGN can remove certain objects from an image, reprocess it, or replace it with a different image like [238]'s replacement of horses with zebras in the famous Cycle-consistent Adversarial Network (CycleGAN).

Figure 3.12 shows an example of the encoder and decoder parts of a DC-IGN with an input x and an output $P(x|z)$ generated using graphic code $Q(zi|x)$.

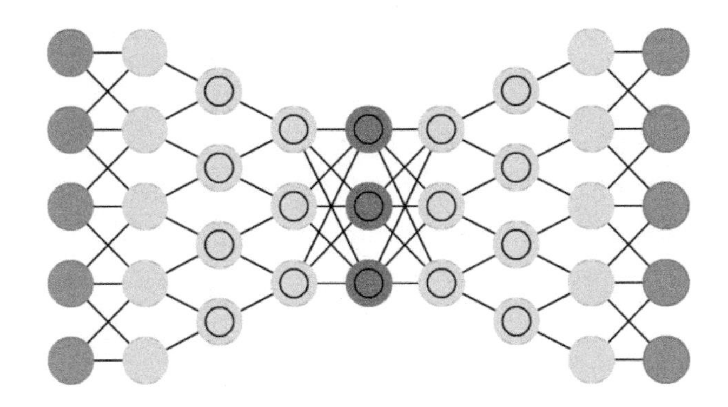

Figure 3.11: Deep Convolutional Inverse Graphics Network (DC-IGN) structure [201].

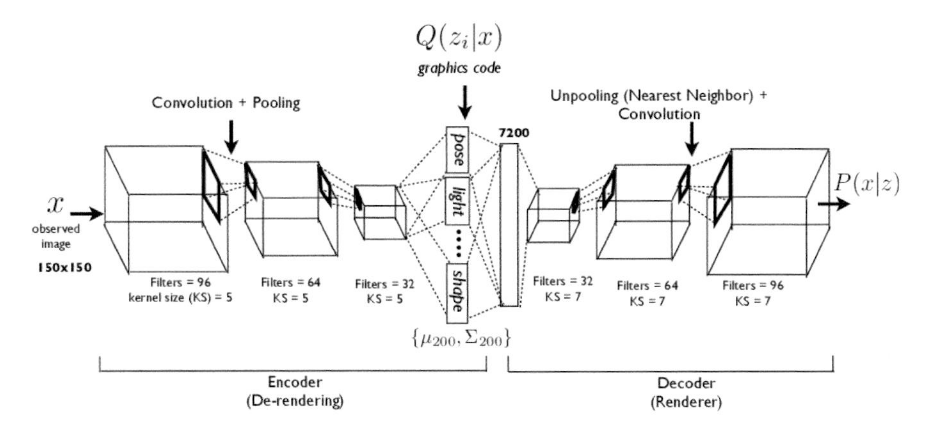

Figure 3.12: Deep convolutional inverse graphics network architecture (Veen, 2017).

3.5 Recurrent Neural Network (RNN)

RNNs are designed to address sequential data. Producing an output at each time step, with recursive connections between hidden units allowing them to maintain a memory of previous data. This makes them well-suited to model time-series [149]. RNN introduces the concept of recurrent cells whereby each hidden cell received its own output with fixed delay. By unfolding one of the RNN states in time as shown in Fig. 3.13, the manipulated input vector u_t and the previous state vector x_{t-1} are utilized as the inputs to the current state x_t, with the weight matrix θ, where x_{t-1}, x_t, and x_{t+1} are all internal states of an RNN model.

As an example of RNN processing, words get transformed into machine readable vectors, then the RNN processes the sequence of vectors in one or more iterations with the input and previous hidden state combined to form a vector. The

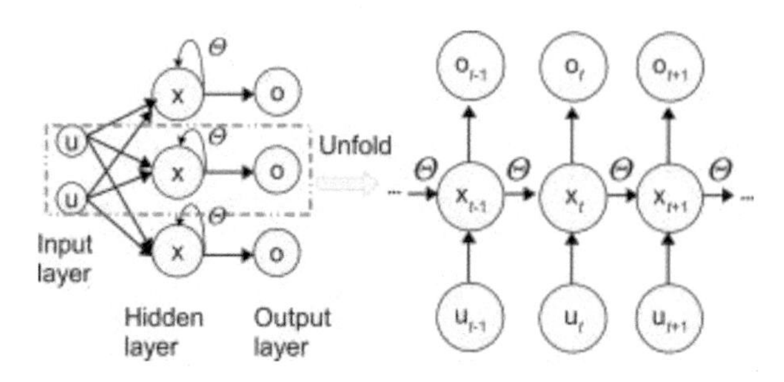

Figure 3.13: General structure of a recurrent neural network and its unfolded structure.

vector, containing information on the current input and previous inputs, goes through the tanh activation, and output a new hidden state, or the memory of the network. Sequential iterations of RNN cells is shown in Fig. 3.14. The RNN passes the previous hidden layer to the next step with the hidden state acting as the NN memory and holding information on the previous data that the network has 'seen' before.

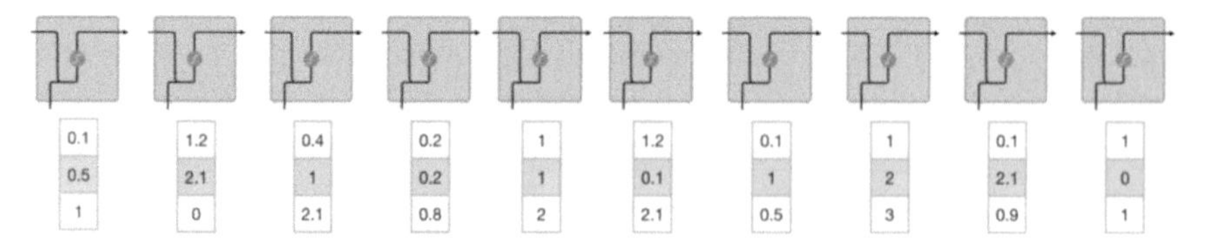

Figure 3.14: Sequential iterations in recurrent neural network.

When vectors are flowing through a neural network, it undergoes many transformations due to various math operations. Under conditions of normal transformations, some values can explode to astronomical levels, causing other values to seem insignificant. To prevent this, tanh activation is used in RNN to regulate the transformation by limiting values flowing through the network to take values between -1 and 1 only. The derivative of tanh is more steep, making it more efficient due a wider range for faster learning and grading. As a result, in certain conditions like short sequences, RNN can use less computational resources than alternative networks.

3.6 Kohonen Network (KN)

Kohonen network (KN), also known as Kohonen map, Self-Organizing Map (SOM) or Self-Organizing Feature Map (SOFM) [114, 115, 167] is a computa-

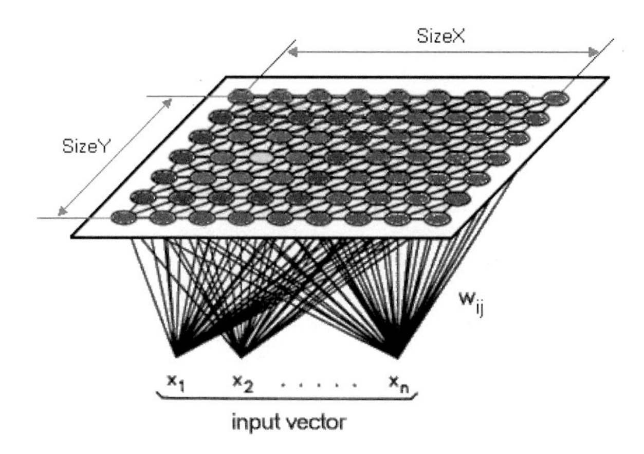

Figure 3.15: Structure of Kohonen network [48].

tional method for the clustering, compression, visualization and analysis of high dimensional data (Kohonen, 2013; Kohonen and Honkela, 2007). Introduced by the Finnish professor Teuvo Kohonen in the 1980s [114, 116, 115], KN is a special case of the class of competitive learning algorithms designed for visualization. KN differs in both its architecture and algorithmic properties from other NNs. Figure 3.15 shows the Kohonen architecture [48].

It is an unsupervised learning model, for applications in which maintaining a topology between input and output spaces is of importance [48]. KN has a single-layer linear 2D grid of neurons, rather than a series of layers and the nodes are not connected to each other but are connected directly to the input vector. As a result, the nodes are only able to update the weight of their connections as a function of the given inputs, since they do not know the values of their neighbors. The grid organizes itself at each iteration as a function of the input such that after clustering, each node has its own coordinates (i, j), and based on this, the Euclidean distance between two nodes can be calculated.

Only a single node is activated at each iteration in which the features of an instance of the input vector are presented to the NN, and all nodes compete for the right to respond to the input. One node is chosen based on the similarity between the current input values and all the nodes in the grid referred. The chosen node, known as the Best Matching Unit (BMU) has the smallest Euclidean difference between the input vector and all nodes. Its position, along with its neighbouring nodes within a certain radius, are slightly adjusted to match the input vector. This is done for all the nodes present on the grid, such that the entire grid eventually matches the complete input dataset, and similar nodes are grouped together, and dissimilar ones separated as shown in Fig. 3.16. KN differs from pure competi-

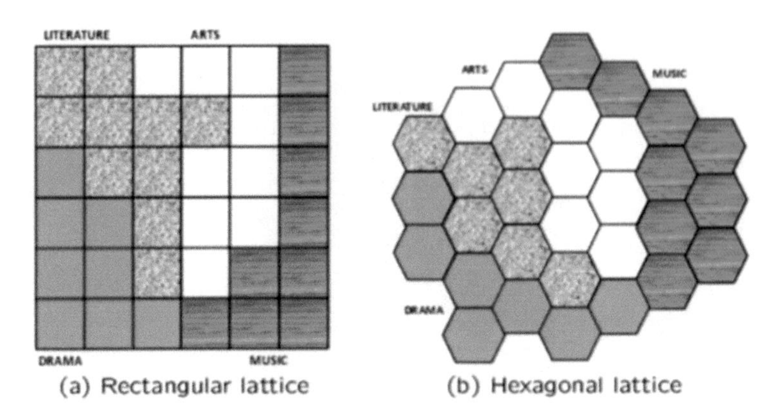

Figure 3.16: Example of visualization of the lattice structure of Kohonen networks [167].

tive learning in that in it imposes a 2-dimensional grid or hexagonal structure on the neurons as shown in the figure, whereas pure competitive learning does not impose any relationship among clusters.

Kohonen network is composed of a grid of output units and N input units. The input pattern is fed to each output unit. The input lines to each output unit are weighted. These weights are initialised to small random numbers. In KN, the correct output cannot be defined a priori, that is, a numerical measure of the magnitude of the mapping error cannot be used. This represents the main difference between KN and conventional networks. The learning process however leads to the determination of well-defined network parameters for a given application. Modifications in KN are such that the weights in the winner neuron are updated in a similar manner to that in vanilla algorithm and the damped version of the update is applied to lattice-neighbours of the winner neuron. In this way, KN also differs from standard competitive learning. The overall effect is that similar points are moved to lattice-adjacent clusters, and the clusters are organized in a 2-dimensional format, which is useful for visualization.

3.7 Deep Residual Network (DRN)

DRN are similar to RNN without explicit delay. Part of input data is passed to next layers, allowing them to be up to about 300 layers deep. It has been empirically shown that there is a maximum threshold for depth with traditional CNN model using convolutional and fully-connected layers typically between 16 and 30 layers [77]. In Fig. 3.17, the plots of training error (left) and testing error (right) against iterations, for 20 layers and 56 layers, show that deeper networks

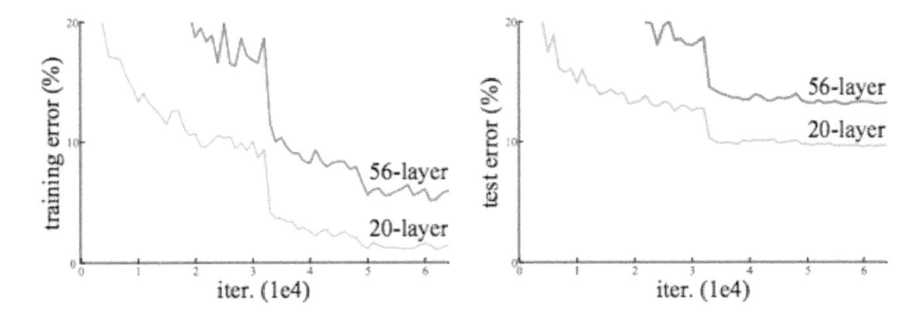

Figure 3.17: Training error (left) and test error (right) on 20-layer and 56-layer CNN [77].

have higher training error and higher test error. A residual is the vertical distance between a data point and the regression line; it is the error that isn't explained by the regression line or the difference between the predicted value $\tilde{y}$ and the observed value. DRN deals with some of these problems by using residual blocks, which take advantage of residual mapping to preserve inputs.

The plot shows that overfitting, as a convenient explanation for the failure of the network, could not be substantiated. In [77], the authors further confirmed that the use of Batch Normalization ensures that the gradients have healthy norms, hence, the failure could also not be blamed on vanishing gradients. Introducing the residual block (a new NN layer), provides an effective means of solving the problem of training very deep networks. A DRN (deep ResNet) supports more sophisticated DL tasks and models. It uses shortcuts or short-circuiting (skip connections) to turn shallow layers to deep layers without degradation in performance, thus enabling the building of very deep networks.

3.8 Long Short-Term Memory (LSTM)

Analysis of error flow in existing RNNs shows that long time lags were inaccessible to existing RNN architectures, as a result of the blow up or exponential decay of back-propagated error. In domains like speech recognition, machine translation, and other complex problem domains, learning order dependence in sequence prediction problems is an important required behavior. In RNN, small gradient updates that halt learning or exploding gradient problems that cause destabilization of the model and inability to learn from training data are both challenges addressed by LSTMs using 'gates', an internal mechanism that regulates the flow of information. LSTMs are a special group of RNN with capabilities for learning long-term dependencies. An LSTM layer is made up of recurrently connected blocks, or memory blocks similar to a differentiable version of the memory chips in a digital computer. The memory cell is a special cell that can process data with time gaps or lags. Each one contains the input, output and forget gates and one or more recurrently connected memory cells. The three gates provide continuous analogues of write, read and reset operations for the cells and the NN is able to interact with the cells via the gates only. Figure 3.18 shows the similarity between LSTM and GRU networks. RNN memories are limited to ten previous words, whereas LSTMs can handle much more sophisticated data, for example, multiple video frames, and are thus widely used in writing and speech recognition.

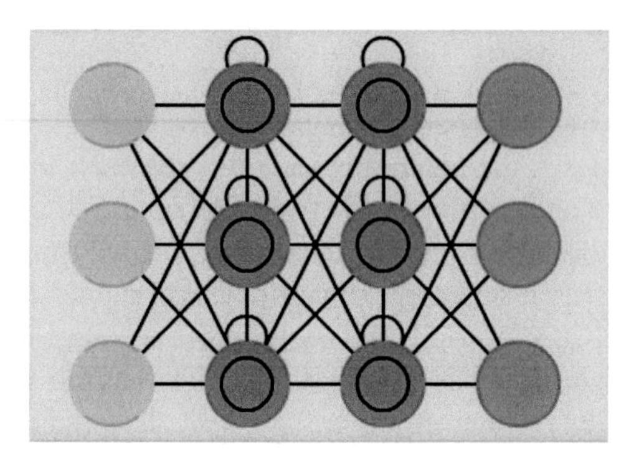

Figure 3.18: LSTM neural networks architecture [201].

Gates in memory cells are recurrent and they control how information is being remembered and forgotten as shown in the peephole LSTM structure in Fig. 3.19, showing the input, output and forget gates with no activation functions between blocks. The gates are able to learn to discriminate between data to determine which one in a sequence should be kept or thrown away. In this way, relevant information is passed down the sequence chain for making predictions. LSTM and GRU are responsible for most state of the art RNN and non-relevant data is forgotten; they have extensive applications in speech recognition and synthesis, and text generation, including generating captions for videos. The LSTM cell has its own memory, and acts as a transport highway that stores and transfers information outside the learning flow of the neural network.

Control flow in an LSTM is similar to that in an RNN; there is a cell, and an input gate, an output gate and a forget gate. Similar to the operation in an RNN, LSTMs processes data in a feedforward manner; its difference lies in the cell operation which enables it to keep or forget information. The cell and gate components enable LSTMs to overcome the limitations of RNNs. Because information from the earlier time steps is able to make its way to later time steps,

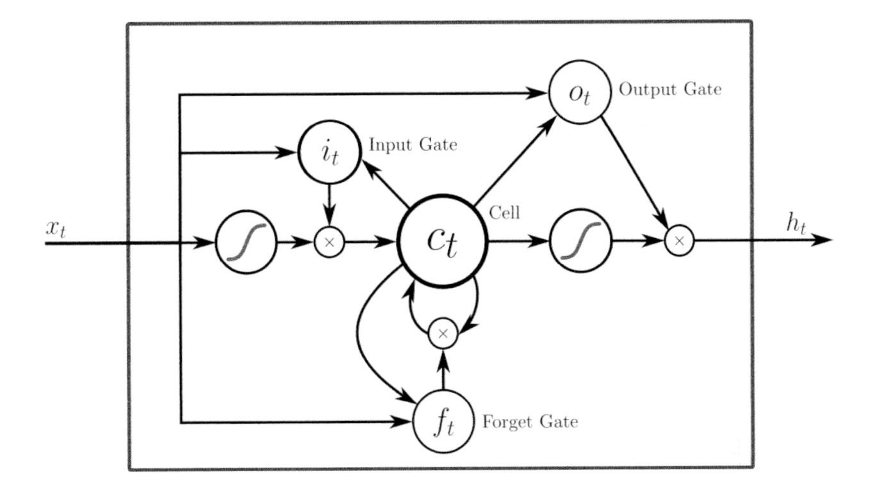

Figure 3.19: A peephole LSTM cell with input, output and forget gates [66].

LSTM is able to address the problems of long-term dependencies by reducing the effects of short-term memory. The LSTM cell corresponds to a node of a recurrent network and has, in addition to the input and output, a forget gate that avoids overfeeding of the vanishing gradient. As the cell state transports information along the sequence, information is added to or removed from it via gates. The gates are able to learn and decide which information is relevant, and thus which to keep or forget during training. A forget gate helps to avoid overfeeding of the vanishing gradient.

Sigmoid outputs a value between 0 and 1; therefore, it can allow zero or complete flow of information throughout the gates. It is used as the gating function for the in, out, and forget gates in LSTM. While forget gate decide what information from previous inputs to forget, input gate decides which new information to remember while output gate decides the output of the LSTM. That is the part of the cell state to output. In the forget gate information from the previous hidden state and the current input pass through the sigmoid function and outputs between 0 and 1. Values of 0 and approximately 0 indicate to forget, while values of 1 or approximately 1 means to keep.

The input gate is responsible for updating the cell state. The previous hidden state and current input are passed into a sigmoid function. The values to be updated are determined by transforming them to be between 0 and 1 where 0 implies not important, and 1 implies important. To regulate the network, the hidden state and current input are also passed into the tanh function where the values are transformed to be between -1 and 1. The tanh and the sigmoid output are then multiplied. The sigmoid output being 1 or 0, decides which information is important to keep from the tanh output. The cell state is calculated using this information. The cell state gets pointwise-multiplied by the forget vector. Values that get multiplied by 0 are dropped. Outputs from the input gate are also pointwise-added. The cell state gets updated to new values that are now relevant for the NN. This produces the new cell state.

The output gate determines the next hidden state which contains information on previous inputs and is used for predictions. The previous hidden state and the current input are passed into a sigmoid function and the newly modified cell state is passed into the tanh function which is then multiplied by the sigmoid outputs to decide the information the hidden state carries. The output is the hidden state. The new cell state and the new hidden state is then carried into the next time step.

3.9 Gated Recurrent Neural Networks (GRU)

Issues of vanishing gradients in DL with feed-forward NNs led to the popularity of new architectures like RNN and new activation functions like ReLUs. For RNNs, skipping the training of the recurrent layers was used to address the problem. Using this method, a simple linear layer is added to the output and the recurrent feedback and parameter initialization is chosen such that the system is very quasi-stable (nearly unstable). The linear layer is the only layer that is then trained. The shortcut procedure enables good performance on many tasks by ignoring the vanishing gradient problem. The limitation of this type of quasi-stable dynamic reservoirs, the effect of any given input can persist for a very long time. The dynamic reservoir must be kept at the quasi stable condition for long-term dependencies to persist, as such, over time, continued stimuli could blow up cause the output. In addition, only the linear layer is learning and there is no direct learning on the earlier parts of the network. Though two decades later, classic LSTM still achieves state of the art results on a variety of sophisticated tasks, modifications to the original LSTM architecture have been suggested and they include the Gated Recurrent Unit (GRU).

A GRU architecture is much simpler than a classical LSTM; the main difference is that the GRU has only two instead of the three gates in LSTM. GRUs combine the gating functions of the input gate j and the forget gate f into a single update gate z, it therefore has only the reset and update gates, whereas LSTM has the input, output, and forget gates. Figure 3.20 shows the basic operations of LSTMS and GRUs.

Because GRU is less complex than LSTM it is the preferred option for tasks involving small dataset while LSTM has better application with larger dataset. Cell state positions earmarked for forgetting are then matched by entry points for new data. In addition, in GRU, the cell state and hidden output h are combined

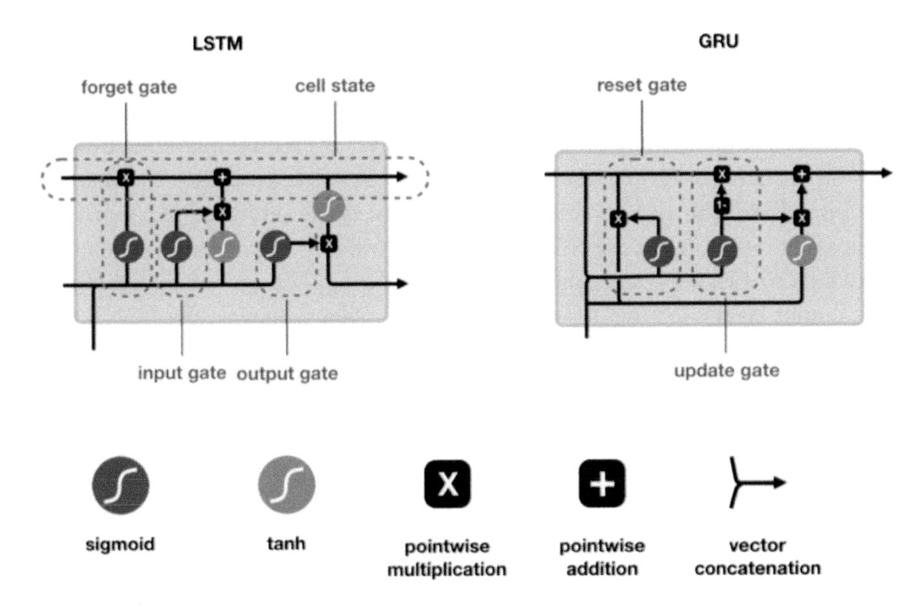

Figure 3.20: The basic LSTM and GRU architecture [150].

into a single hidden state layer, while the unit also contains an intermediate, internal hidden state.

GRUs have shown great capabilities for both sophisticated tasks like Neural GPUs as well as simpler ones like machine translation or music and text generation, they are however limited in tasks like counting, and as such, remains less powerful than classic LSTMs.

Other architectures based on the LSTM include the multiplicative LSTMs (mLSTMs) by [117] and LSTMs with Attention which is regarded as probably the most transformative innovation in sequence models. Attention is a model's ability to focus on specific elements in data (for example, the hidden state outputs of LSTMs in this case). It was employed by [218] as a sandwich between encoding and decoding LSTM layers to achieve state-of-the-art Neural Machine Translation. Google Translate was also powered by attention LSTM. Another example of the capability of the architecture is OpenAI's demonstration of tool use in a hide-and-seek reinforcement learning environment. Other architectures include the memory-hungry transformer [8] and many language models like Bert, ELMo and GPT-2 [44, 53]. Several LSTM variants have been reviewed for performance on various typical tasks and many, including simulated mutants, have been found to have capability greater than, or similar to those of classic LSTM and GRU in some of the tasks studied [95].

3.10 Bidirectional Recurrent Neural Network (BRNN)

One shortcoming of conventional RNNs is that they are only able to make use of previous context. Bidirectional RNNs (BRNNs) overcome this limitation of RNN by utilizing both previous and future contexts. It processes the data in both directions with two separate hidden layers, which are then fed forwards to the same output layer. BRNNs present each training sequence forwards and backwards to two separate recurrent nets, both of which are connected to the same output layer such that for every point in a given sequence, the BRNN has complete, sequential information about all points before and after it. Combining BRNNs with LSTM results in bidirectional LSTM, which can access long-range context in both input directions. The basic structure of a BRNN is shown in Fig. 3.21.

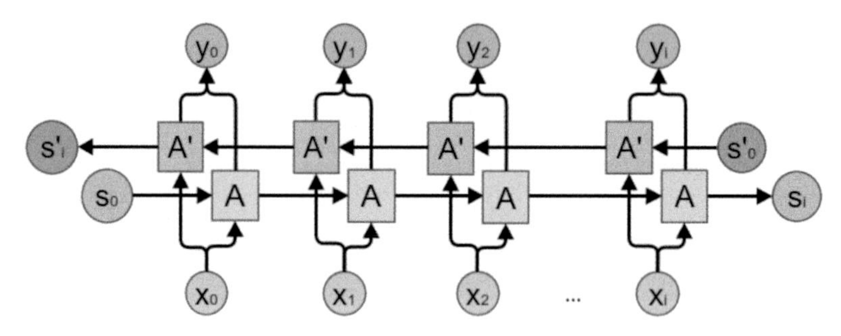

Figure 3.21: The bidirectional recurrent neural network architecture.

3.11 Hopfield Network (HN)

Invented by Hopfield in 1986, HN is an RNN with symmetric connections. It is a fully connected network in which every neuron's output is an input to all the other neurons. The Hopfield network is not a general RNN; it is not designed to process sequences of patterns and unlike in the one-directional, multilayer perceptron and input-output mapping network type of RNN, in HN, neurons connect to other neurons, forming single or multiple feedback loops as shown in both the left and right illustrations in Fig. 3.22. The HN neurons do not have self-loops but are fully connected, resulting in $K(K-1)$ interconnections if there are K nodes, with a w_{ij} weight on each such that the neurons transmit signals back and forth to each other in a closed-feedback loop, and thus settling in stable states.

Hopfield networks are trained on a limited set of samples so they respond to a known sample. Each cell serves multiple purposes as input cell before training, as hidden cell during training and as output cell when used. HNs attempts

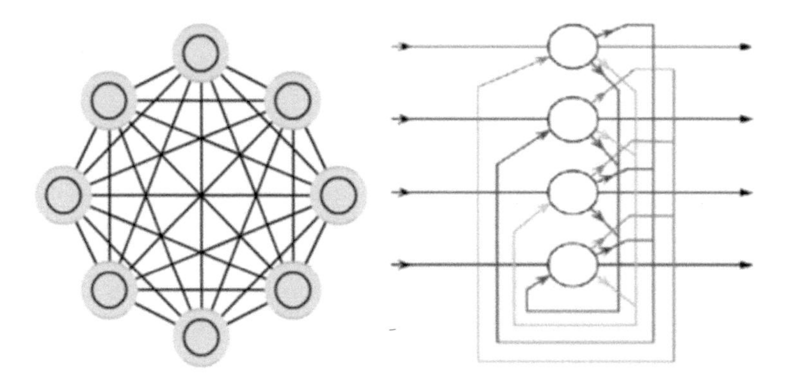

Figure 3.22: Structure of the Hopfield network.

to reconstruct the trained sample, and can therefore be used for denoising and restoring inputs. If the network is provided with half of learned picture or sequence, it will return a full sample. HN is based on the assumption of symmetric weights for neural interactions, that is, $w_{ij} = w_{ji}$, which is unrealistic since in real neural systems, this is an unlikely condition. The state s_i of a unit is either $+1$ or -1.

The rule of activation is given by:

$$s_i = \begin{cases} +1 & if \sum_j w_{ij} s_j \geq \theta_i \\ -1 & otherwise, \end{cases}$$

where θ_i is a threshold value corresponding to the node. This activation function mirrors that of the perceptron. A node is picked to start the update, and consecutive nodes are activated in a pre-defined order. In synchronous mode, all units are updated at the same time. In other models, for example, the Hebbian learning model, neurons are activated simultaneously, such that there are increments in synaptic strength between the neurons and the higher the weight w_{ij}, the more likely it is that connected neurons will activate simultaneously. Hebbian learning in HN is captured as:

$$w_{ij} = \frac{1}{N} \sum_{k=1}^{N} x_{ki} x_{kj}$$

Here x_k is in binary representation — that is, the value x_{ki} is a bit for each i. If the connections are trained using Hebbian learning then the Hopfield network can perform as a robust content-addressable memory, resistant to connection alteration. Hopfield networks have a scalar value associated with each neuron of the network that resembles the notion of energy. The sum of these individual scalars gives the "energy" of the network as shown in the equation:

$$E = \frac{1}{2} \sum_{ij} w_{ij} s_i s_j + \sum_{i} \theta_i s_i$$

The value of E decreases or remains the same if network weights are updated for learning. The interaction term also resembles the Hamiltonian of a spin glass. A HN is an associative memory, hence, its operation differs from that of a perceptron, which is a pattern classifier. For example, in a hand-written digit recognition task like multiple examples of a digit written in various ways. A HN associative memory would recall a canonical pattern for the digit that was previously stored rather than just classifying the digit directly. In this way, an associative memory presents as a form of noise reduction. This associative memory has a storage capacity determined by the learning model. For example, with Hebbian learning, it is about $N \leq 0.15K$ whereas the quantum variant of Hopfield networks provides an exponential increase over this. Similar to BRNN or bi-LSTM, HN variant that implements the bidirectional approach, the bidirectional associative memory (BAM) features two layers, each of the layers can serve as an input, and produce an output on the other layer. Apart from HNs other known RNNs are fully reconnect, recursive, Elman, and Jordan among others.

3.12 Generative Adversarial Networks (GAN)

Generative Adversarial Networks (GANs) are a model architecture for training a generative model. It was proposed by [65] based on the simultaneous training of a generative model and a discriminative model. The generative model, G, captures data distribution, while the discriminative model D estimates the probability that a sample came from the training data rather than the generative model. The framework corresponds to a minimax two-player game. In the game scenario in GANs the generator network competes against the discriminator network, which represents the adversary [63]. While the generator network directly produces samples, the discriminator network tries to distinguish between the 'fake' samples from the generator and those from the training data. GAN thus represents a huge family of double networks that attempts to fool each other. The discriminator is updated with successive effort at distinguishing between fake and real samples, and thus gets better at in successive rounds. The generator also gets updated on how well, or not, the generated samples were able to fool the discriminator.

In 2015, [158] formalized the more stable GAN approach, the Deep Convolutional Generative Adversarial Networks (DCGAN) through which more stable models were developed. They provided the following architecture guidelines for stable DC-GANs:

- Replace any pooling layers with strided convolutions (discriminator) and fractional-strided convolutions (generator).
- Use batchnorm in both the generator and the discriminator
- Remove fully connected hidden layers for deeper architectures.
- Use ReLU activation in generator for all layers except for the output, which uses Tanh.
- Use LeakyReLU activation in the discriminator for all layers.

Generative modeling is an unsupervised learning problem, as we discussed in the previous section, although a clever property of the GAN architecture is that the training of the generative model is framed as a supervised learning problem. Based on tools of game theory, the competition between the two models in a GAN framework qualifies them as adversaries playing a zero-sum game. The idea of zero-sum game refers to the fact that the discriminator is rewarded when it successfully distinguishes between the samples, and no change is made to the model parameters, whereas the generator gets penalized with large updates to model parameters. The exact opposite also holds with the generator rewarded when it successfully fools the discriminator, and the discriminator penalized. Figure 3.23 shows the structure of the GAN model and Fig. 3.24 shows the operations of the generator and the discriminator.

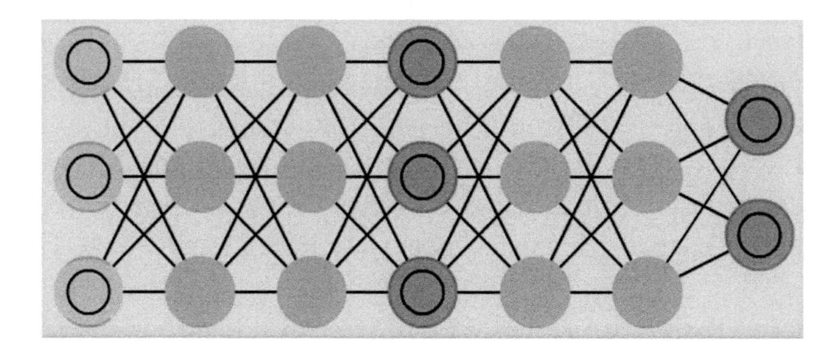

Figure 3.23: The generative adversarial network architecture [201].

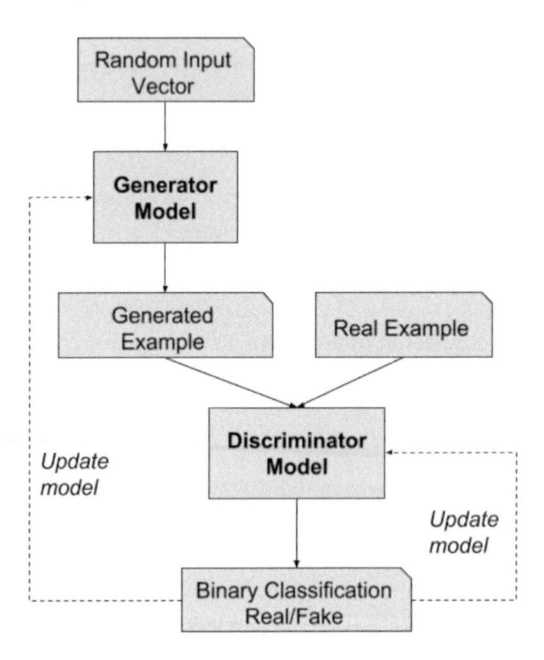

Figure 3.24: Structure of the generative adversarial network showing the operations of the generator and the discriminator.

3.13 Deep Belief Network (DBN)

Deep Belief Networks (DBNs) are another group of generative, stochastic NN proposed as solutions to multiple training problems with traditional NN training in deep layered networks [6]. DBNs are made up of stacked modules of restricted Boltzmann Network (RBM) [37] with two layers of feature-detecting visible (v) and hidden (h) units, or a special type of Markov random field [164]. A DBN features connections only between layers, and can learn a probability distribution from its input datasets by searching the parameter space of deep architectures and thus address issues like the need for huge training dataset and slow learning, among others. DBNs differ from conventional shallow learning NNs, as they can facilitate identification of deep patterns. They possess deep reasoning abilities and are therefore able to distinguish between normal and faulty data. The general structure of a DBN is shown in Fig. 3.25.

Each stacked RBM (that is, RBM1, RBM2, RBM3, etc.) in the DBN architecture is made up of a visible layer v and a single hidden layer h_n. The first RBM is trained using the input data as visible units. The second hidden layer h_2 of the second stack is trained using the output of the previous trained layer h_1. The out-

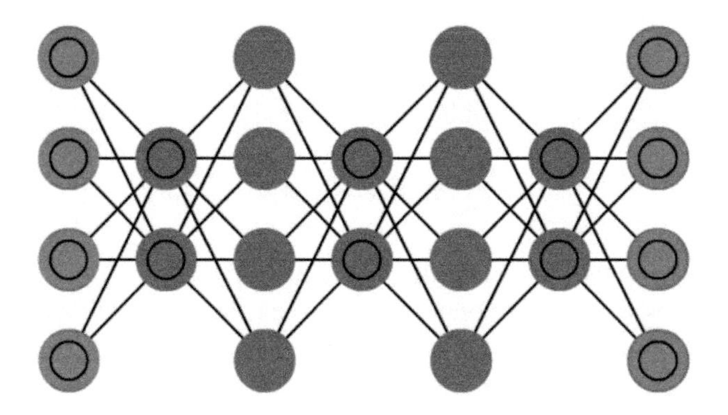

Figure 3.25: Structure of the deep belief network [201].

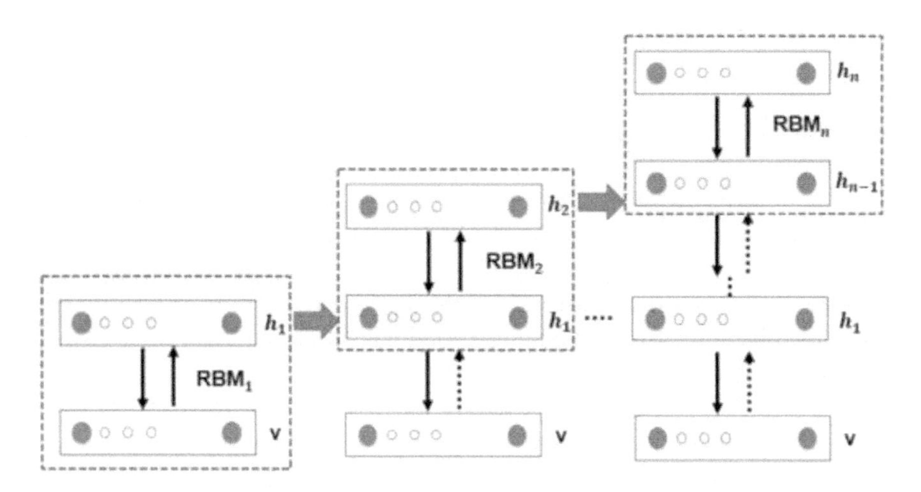

Figure 3.26: Deep belief network (DBN) architecture composed by stacked RBMs.

put of h_2 becomes the input of the next RBM3 and so on. The trained layers form the stacked architecture as shown in Fig. 3.26. The whole DBN is tuned with a standard back propagation algorithm. DBNs have shown higher accuracy, with extensive applications in image recognition, speech recognition, hand-writing recognition and other classification problems compared with some existing DL methods. Unsupervised learning is implemented in each RBM module using contrastive divergence procedure. The features learned (output) from one stage is used as input the subsequent after which supervised learning is implemented at the whole network level to improve classification performance. This procedure is referred to as a fine-tuning method.

The stacked RBM in a DBN, is a stochastic recurrent neural network with two layers of visible units, v, and a layer of binary hidden units, h [137]. The total network energy of the visible and hidden units (v,h) is given by:

$$E(v,h) = \sum_{i,j} v_i h_j W_{ij} - \sum_i v_i a_i - \sum_j h_j b_j$$

where i represents the indices of the visible layer, j those of the hidden layer, and $w_{i,j}$ denotes the weight connection between the i_{th} visible and j_{th} hidden unit. Further, v_i and h_j denote the state of the i_{th} visible and j_{th} hidden unit, respectively, and a_i and b_j represent the biases of the visible and hidden layers. The first term, $\sum_{i,j} v_i h_j W_{ij}$ represents the energy between the hidden and visible units with their associated weights. The second, $\sum_i v_i a_i$ represents the energy in the visible layer, while the third term, $\sum_j h_j b_j$ represents the energy in the hidden layer. The RBM defines a joint probability over the hidden and visible layer $p(v,h)$, that is:

$$p(v,h) = \frac{e^{-E(v,h)}}{z}$$

where Z is the partition function, obtained by summing the energy of all possible (v,h) configurations, $Z = \sum_{v,h} e^{-E(v,h)}$.

To determine the probability of a data point represented by a state v, the marginal probability is used, summing out the state of the hidden layer, such that $p(v) = \sum_h p(v,h)$. Using the equation, the probability of either of the visible or the hidden layer being activated can be calculated for any given input. The conditional probabilities in the model can also be determined using these figures. To maximize the likelihood of the model, the gradient of the log-likelihood with respect to the weights must be calculated. The gradient of the first term, is given as:

$$\frac{\delta log \sum exp(-E(v,h))}{\delta W_{ij}} = v_i . p(h_j = 1|v)$$

Although the gradient of the second term cannot be computed, the hidden and visible layers can be computed using the following two equations:

$$p(h_j = 1|v) = \sigma(b_j + \sum_i v_i W_{ji})$$

and

$$p(V_i = 1|h) = \sigma(a_j + \sum_j h_i W_{ji})$$

where $\sigma(.)$ is the sigmoid function.

The learning rule which performs the steepest ascent in the log probability of the training data is given by:

$$\frac{\delta log(\sum_h(-E(v,h)))}{\delta W_{ij}} = \langle v_i h_j \rangle_0 = \langle v_i h_j \rangle_\infty$$

where, $\langle . \rangle_0$ and $\langle . \rangle_\infty$ are the respective expectations for the data distribution (p_0) and model distribution.

A DBN is composed of an arbitrary number of stacked RBMs, resulting in a mix between a partially directed and partially undirected graphical model, such that overall distribution between visible layer v (input vector) and the l hidden layers h^k is defined by:

$$p(v, h^1, ..., h^k) = \prod_{k=0}^{l-2} P(h^k|h^{k+1}) P(h^{l-1}, h^1)$$

where $P(h^k|h^{k+1})$ represents a conditional distribution for the visible units conditioned on the hidden units of the RBM at level k, and $P(h^{l-1}, h^1)$ is the visible-hidden joint distribution in the top-level RBM.

Chapter 4

Top Applications of Deep Learning Across Industries

DL allows machines to solve relatively complex problems even by using diverse and less structured data. That is why, nowadays, DNNs are used on a variety of tasks along various industries such as healthcare, defense, education, etc. In the corresponding sections we will review top applications of DNNs across various industries.

4.1 Agriculture

AI is expanding its presence in all industries, including even the world's most vital domains such as agriculture. With the help of cutting-edge technologies, it helps to achieve better productivity and crop yielding. Agriculture and farming play a key role in the economic sector. To fulfill the demand of global population by 2050, which will reach more than nine billion, agricultural production must be dramatically increased. To produce more food to feed a growing community, AI-driven technologies can help address the challenges the agricultural domain is facing nowadays, including labor shortages, market demand predictions, effective treatment analysis, and more.

Agricultural work is difficult, and labor shortages are one of the hot topics in the industry. Farmers are trying to solve this problem with the help of agricultural robots, which are predicted to complete an increasingly diverse range of tasks by 2030. The agricultural sector has already begun to benefit from intelligent spraying systems, driverless tractors or AI-powered robots for harvesting. These applications allow to get work done faster and more accurate without hiring more people. For example, Harvest CROO Robotics has developed a robot called Sweeper that helps collect and pack sweet peppers replacing 30 human laborers [11]. This robot can achieve an accuracy of about 61% when harvesting ripe peppers. In [229], the authors proposed the Regions-CNN method to automatically detect branches on apple trees. This can be integrated into the automation of the shaking process to avoid manual picking of vegetables and fruits from the branches, which is becoming an increasingly difficult task for farmers. Another fast and accurate DL-based model has been suggested in [144]. They used color camera and a CNN-based Single Shot MultiBox Detector (SSD) to locate the fruit (Fig. 4.1).

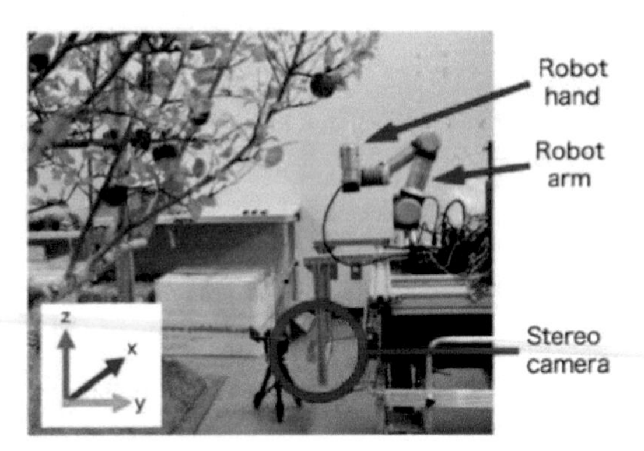

Figure 4.1: Harvest robot [144].

Pest infestation control is also important for farmers to avoid crop damages globally prior to harvest time. For a long time, chemical farming was the only solution to avoid crop loss due to diseases or pests. However, the chemicals present in food have a very bad effect on living organisms. That is why it is necessary to find another way to detect and eliminate pests or diseases in their early stages without harming crops. DL algorithms can detect the location of various insects using satellite images. Typically, image processing techniques are used to determine leaf damage, followed by pest classification using NNs [189]. Such approaches allow to identify pests in their early stages, so the amount of pesticides can be intensively reduced. Acoustic detection mechanisms are another accurate approach for identifying various types of pests [185]. Figure 4.2 shows a model for acoustic identification of pests in cauliflower crops. The authors have proved that the efficiency of pest detection and elimination in the crop is approximately 98.9% without harming the crop itself.

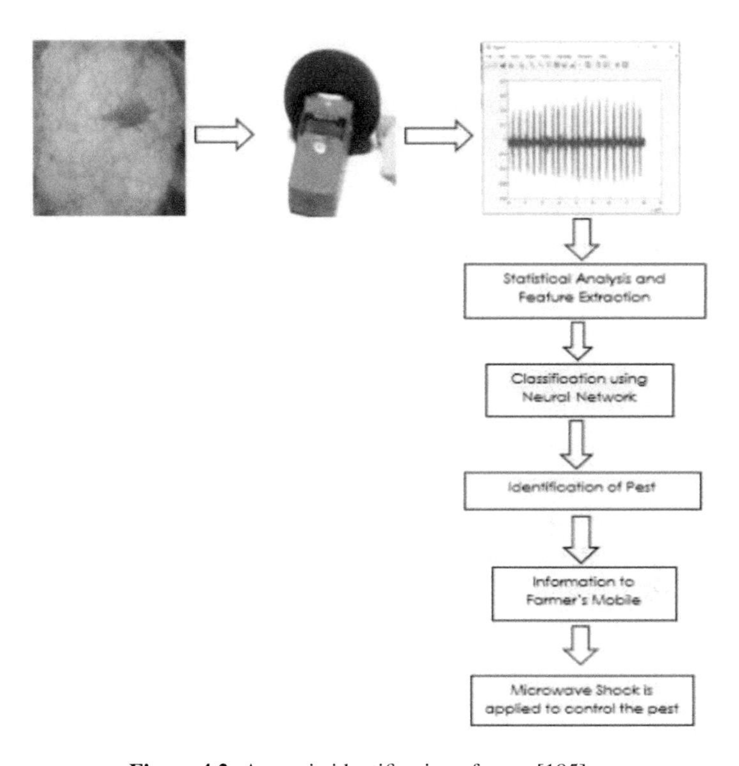

Figure 4.2: Acoustic identification of pests [185].

To protect and improve crop yield, the agriculture domain relies heavily on AI solutions. By combining these approaches with IoT devices, farmers can get more accurate information in real time, which means less time and costs spent on trial and error. Consequently, companies are leveraging DL algorithms to process data captured by IoT devices to monitor the health of crops and soil. Due to soil or other conditions, diseased crops should be eliminated immediately. For example, AgEYE Technologies develops ML-based sensor solutions to detect pathogens and contamination in a fast manner (https://ageyetech.com/). Blue River Technology has developed another robot called See and Spray, which is a

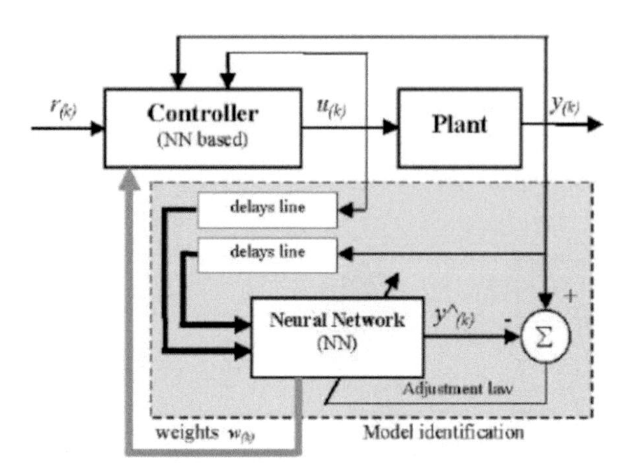

Figure 4.3: NN based controller [26].

precision spraying solution. They use DL algorithms similar to face recognition to detect and classify the types of weeds (https://bluerivertechnology.com/).

Smart irrigation systems can help save wasted water, an important concern for farmers. IoT devices can help here control water pumps based on ML/DL algorithms [96]. In [26], the authors proposed a NN-based controller that regulates the moisture level in the root zones (Fig. 4.3). The close-loop autonomous controller calculates the length of irrigation based on moisture level set by the user and then it predicts the system response. NN is re-trained after each irrigation period and is used to estimate future moisture levels.

Deforestation and degradation can have destructive impacts on the economy. Based on DL algorithms, Plantix is designed to identify potential defects and nutrient deficiencies in soil, helping to address a major challenge in agricultural domain. The company claims that its software can achieve pattern detection with an estimated accuracy of up to 95%. This Android mobile application also offers treatments based on images sent by users (https://plantix.net/en/).

Crop yield prediction is an important but challenging task due to the multiple and complex factors on which it depends, such as environment or genotype. This type of prediction is useful not only for farmers when making financial decisions, but also for policy makers when making profitable import and export decisions. For example, in [102], the authors proposed a DNN-based solution to predict yield, check yield (average yield across all hybrids) and yield difference of corn hybrids. Figure 4.4 shows the DNN architecture for crop yield prediction. The suggested method was able to learn nonlinear relationships between genes, environmental conditions and their interactions based on historical data in order

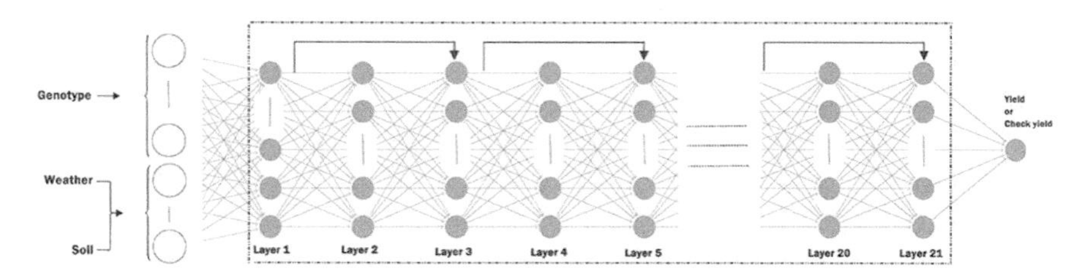

Figure 4.4: Acoustic identification of pests [102].

to make reasonably accurate crop yield predictions. It is also possible to predict harvest area and production using DL methods [1].

Managing grocery needs and educating farmers to grow only the required crops are very important in the agricultural sector of every country. PECAD is a DL algorithm that predicts the future production prices based on past pricing and volume patterns [69]. It consists of two CNNs for pricing and volume data, and achieves significantly lesser root mean squared error. Food market demand prediction is also important for enterprises when organizing an appropriate purchase, production and sales plans. For example, to predict the monthly sales of health food, a back propagation NN and the Particle Swarm Optimization algorithm can be used [87].

AI in agriculture is not only helping farmers automate their farming, but it is also shifting towards precision farming for higher crop yields and better quality while using fewer resources. Agriculture is one sector that is not yet fully based on AI technologies, and there are not yet many applications currently used in this domain. Farmers tend to perceive AI as something that only applies to the digital world, and they may not see how this can help them in their work on the physical land. Their resistance is due to a lack of understanding of the practical application of AI tools. AI is just the next step in smart farming, and to reap all the benefits of AI, farmers will first need technological infrastructure. This infrastructure will take some time to be developed, but then farmers will be able to build a solid technology ecosystem that will stand the test of the time. Below are the main directions of agricultural domain, in which DL solutions are already used somehow:

- Driverless tractors,
- Smart spraying and irrigation tools,
- Recommendations for effective treatment,

- Prediction of crop yield,
- Market demand prediction and analysis.

4.2 Banking and Finance

The banking sector has been using ML algorithms for a long time to drive business growth based on the lower costs, improve accuracy and customer experience. The credit card fraud detection [61], trading models based on forecasting [109] or oil price prediction [58] are some of the popular ML approaches that are actively used in the financial industry. However, the DL became more popular in recent years and more and more DL models started to be used for finance. Hence, in this section, we will thoroughly describe the state of the art DL based applications in the banking sector.

The recent financial crises around the world have brought the need for crisis prediction that will provide early warnings to regulatory agencies. On the other hand, every financial institution, including investors in banking stocks, tries to find out the probable crisis in advance in order to take corrective actions. Consequently, there has been a tremendous need to develop applications for such predictions. In [92], the authors present a new model based on the multilayer perceptrons and self-organizing maps to predict crisis for US banks. Another major event in the financial sector is currency crises affecting the economies of countries. In [7], the authors used deep neural decision trees to predict currency crises in the global market which is a big step towards establishing stability at the international level.

One area in the banking system that has been greatly affected by NNs is related to loan applications. In order to reduce the risks of failed loan applications, recently, many banks started to use NNs for decision making stating that based on accuracy it significantly overcomes human-methods. These systems are usually based on the analysis of past failures and experiences. As a result, different research works have been conducted to find the best NN model for this task. The knowledge discovery tool is an example for this kind of decision context [74]. It consists of three NN models: a MultiLayer Perceptron, an Ensemble Averaging committee machine and a Boosting by Filtering committee machine (Fig. 4.5). The final experimental results conclude that Boosting by Filtering Committee Machine models outperformed other models. Overall, $1.65\% - 2.38\%$ average percentage error on all data confirms the suitability of NNs usage for evaluation of loan applications.

Due to increased capabilities of current systems, NNs play a significant role in image and character recognition. Due to the increased threat of cybercrime in

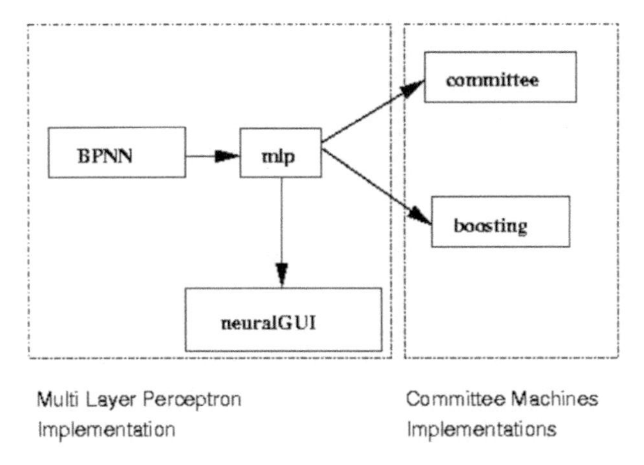

Figure 4.5: Knowledge discovery tool [74].

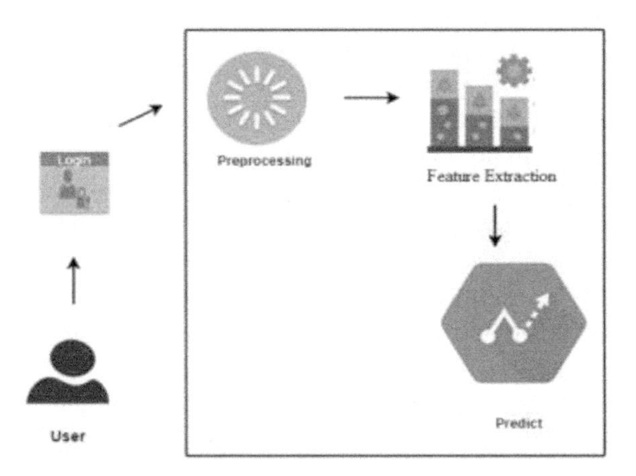

Figure 4.6: Credit card fraud detection system [161].

the banking sector, this advantage is used to better detect frauds such as written signature or credit card frauds. Based on the recent years increased number of online transactions, it is also very important to detect fraud banking accounts and transactions. In [130], CNN models are discussed in detecting fraudulent bank accounts. It is proven that ANNs usage is more suitable for credit card fraud detection giving accuracy more than that of the unsupervised learning algorithms [161]. The authors present the credit card fraud classification and come up with a system to predict the fraud transaction (Fig. 4.6).

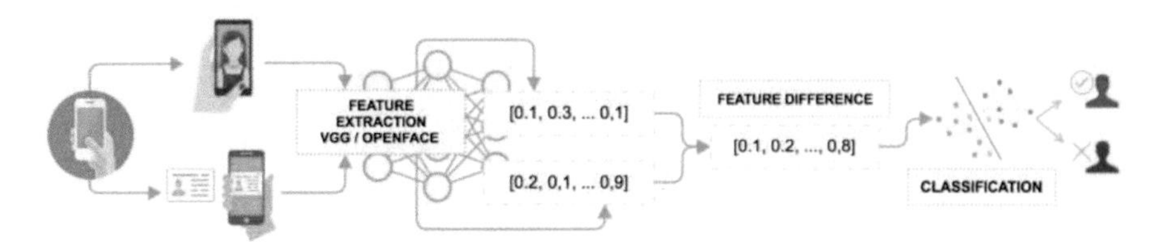

Figure 4.7: The proposed architecture for cross-domain face matching [143].

In addition, to increase security and enhance the customer experience, banking systems use facial-recognition technologies to create a map of a user's faces. This helps to protect customers' data, reduce fraud and prevent money laundering. Currently, the face recognition and authentication approaches are mostly based on CNNs. In [143], the authors suggested a new approach where the feature vectors are extracted by two well-referenced CNNs, VGG-Face and OpenFace (Fig. 4.7). It achieved a high accuracy of 93% making it applicable for the banking sector.

Portfolio management is a process of managing and overseeing a group of investments to maximize the earnings within a given time period. The entire process is based on the ability to make correct decisions. Hence, there are a lot of studies to build a system to achieve this goal. In [239], the authors suggest a portfolio management system, where the portfolio optimization algorithm is based on the FNN and the forecasting approach is based on the error correction NN. They proved it as a promising framework for the financial market. In the last two decades, with the introduction of electronic online trading platforms and frameworks, the trading part of portfolio management has become a most popular area in the finance industry. Since the usage of DNNs allows analyzing thousands of data sources at the same time, the trading strategies can become much more profitable based on these approaches. Hence DL is also used for determining the optimal buy and sell time in the stock market. In [32], the authors showed that the stock trading using the generalized regression NN outperforms the results of stock trading without NN usage. Using a different model, [230] the authors suggested a new State Frequency Memory(SFM) RNN for stock price prediction with multiple frequency trading patterns that achieved better trading performance.

Some of the trading studies are concentrated on the NN models' usage to handle predictions for both stock market indexes and stock values. Based on the NASDAQ stock exchange rates, it was discovered that for nine working days a feed forward ANN with 20-40-20 neurons is the optimized one with higher accuracy [138]. In [233], the authors proved that the DBN model is better compared to the FNN model for exchange rate forecasting. On the other hand, the gated RNN architectures such as LSTM and GRU outperformed traditional RNNs for the same problem [42]. In stock market and exchange rate prediction, some researches show that market news affects the final price changes, hence some filtering needs to be done to extract useful information. For this purpose, a hybrid model is suggested that combines a generative model and DNN to predict the movement of stock prices. Several studies show that these types of hybrid models have better performance in stock and exchange rate prediction models [160].

Recently, more credit card companies and online shops started to use NN solutions for finding the ideal customer to generate sufficient revenue. Market leaders are those who can accurately predict the revenue and risk for each potential customer with low expenses. The DL solutions help to identify people among the general public who most resemble the profiles of existing customers. In addition, these approaches are based on the location data to help increase the benefit of the models. For online shops, one more additional step is taken to avoid unnecessary ads for people who are not going to buy the product [235].

There is a huge number of data and information that is communicated over the Internet each day. That has led to a need of extracting useful information automatically from this unstructured data. One of the crucial elements in the financial sector is text mining that aims to extract useful patterns from textual data. Hence, ML and DL algorithms are used to process the available information. In [70], the authors have done a thorough review of text mining applications in the finance sector. They presented the areas of the financial domain where text mining can be used (Fig. 4.8).

As the area of text mining in general is a very crucial topic with different approaches, we will discuss it as a separate topic in the last section. It is also important to note that in the financial sector as in other industry sectors there are a lot of automation processes and advisory systems based on NNs. It can include call center automation, paperwork automation, financial products recommendation to potential clients, etc.

Hence, the finance industry is one of the most influential industries that has been impacted by new findings in AI. Below are the important areas of banking sector that are mostly leveraged from DL solutions:

- Crisis prediction,
- Decision making for loan applications,
- Fraud detection and security,
- Stock market and exchange rate forecasting,

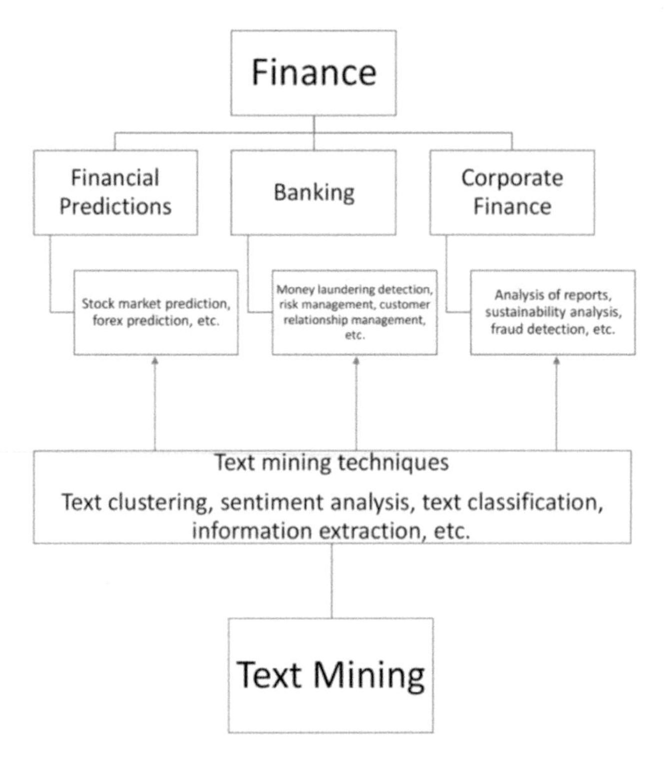

Figure 4.8: Text mining usage in financial domain [70].

- Portfolio management and trading strategies,
- Ideal customer detection and risk management,
- Financial text mining.

4.3 Education

With the early release of Lego Mindstorms kits, developed at the MIT Media Lab in the 1980s, robots have remained popular in education ever since [111]. Intelligent Tutoring Systems (ITS) utilize a variety of DL methods to match students with interactive machine tutors in science, math, language, and other disciplines. Natural Language Processing (NLP), especially when combined with DL and crowdsourcing, has boosted online learning and enabled teachers to multiply the size of their classrooms by magnitudes while simultaneously addressing individual student learning needs and styles. Data sets from large online learning systems have fueled the rapid development of learning analytics for future research.

DL methods in the educational field, especially in teaching, began to spread since 2010. In [24], the authors have proposed a method that effectively classifies student achievements in a particular module. DNNs become useful in assessing the knowledge and skills learned by students during courses. In addition, today the use of NNs in education is the easiest way to detect cases of plagiarism, thereby improving existing approaches to grading and evaluation [51].

Education Data Mining (EDM) is designed to detect patterns in large collections of educational data. It leverages ITS, Learning Management Systems (LMS) and Massive Open Online Courses (MOOC) to obtain information from student learning activities. Discovering the hidden patterns among this information is very useful in the educational sector, as it permits to understand and create more adaptive and personalized offerings for students and tutors. DNNs and their capability to detect previously unexpected connections are very promising tools in this endeavor that can analyze this volume of data to address various educational problems [80].

Promising applications of EDM are student performance and dropout forecasting. For example, for universities in the educational sector, it is important to predict the list of students at high risk of dropout based on several criteria. Such prediction model was proposed in Fig. 4.9 to reduce the number of students at high risk.

The authors identify critical factors that can lead to academic failure and dropout, such as student loan application, number of absences, number of alerted subjects and academic performance. The DL method achieved approximately 90% accuracy compared to logistic regression, which has an accuracy of 88%. This is a good start for future applications to achieve higher accuracy in student performance and dropout prediction.

Assessing academic performance of students is difficult due to diverse factors on which it depends. In [68], the authors presented the Students Performance Prediction Network (SPPN) based on a six layer NN. An overview of the proposed system is shown in Fig. 4.10. A sparse auto-encoder is used for training the hidden layers of features, after which supervised learning is used for fine-tuning the parameters. To achieve better performance, the authors also implement the training process on the Graphical Processing Units (GPU) that overcomes the performance of the Central Processing Unit (CPU) 9 times.

K-12 schools use more sophisticated and versatile kits available from a number of companies that are building robots with new sensing technologies programmable in a variety of languages. For example, Ozobot is a robot that teaches children to code and reason deductively. This teaching method is integrated by

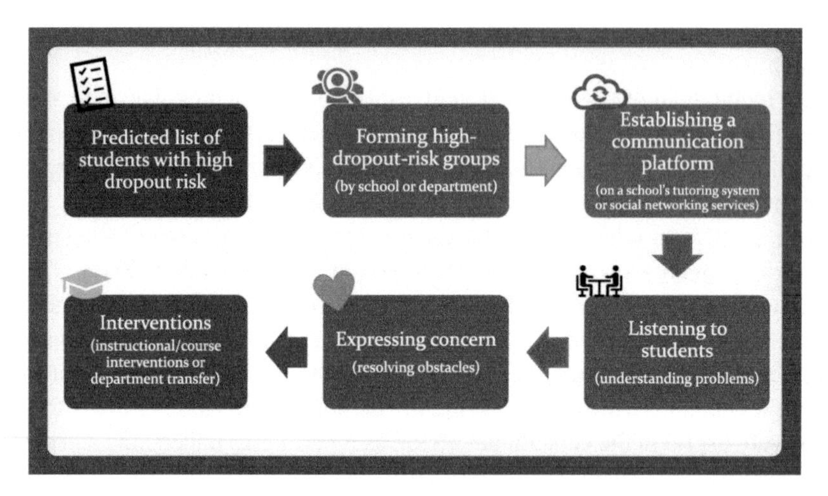

Figure 4.9: Administrative measures to reduce the number of students at high risk group [197].

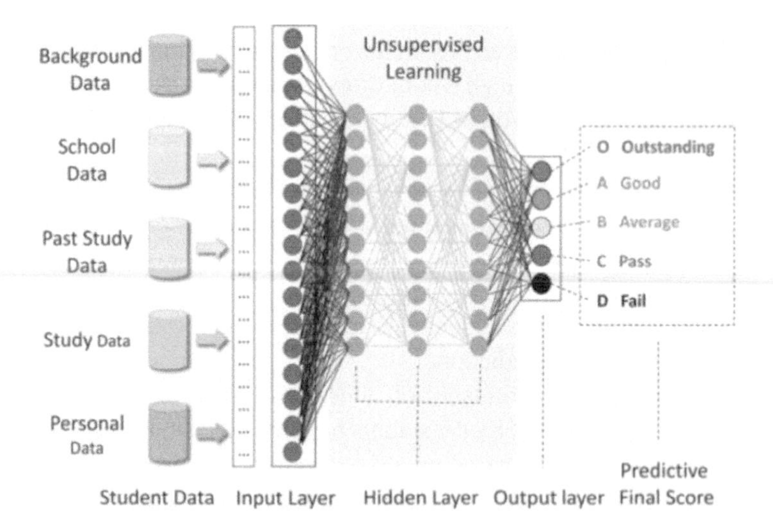

Figure 4.10: Students performance prediction system [68].

configuring it to dance or play based on color-coded patterns. It has been proven that Ozobot is suitable not only for children, but also for the older generation [57]. Cubelets aims to teach children logical thinking through assembling robot blocks so that they can think, act, or sense, depending on the function of the different blocks [212]. Consequently, in recent years, the field of teaching robots has become very active for solving various educational problems. DL started to be used in teaching assistant systems to achieve better results in terms of extracting efficient learning features from existing big data, predicting performance of students or supporting personalized and adaptive learning paths. For example, distraction during online lessons is a common problem in the current educational system. DL methods help to detect the distraction of students to have more intelligent teaching assistants [237]. ML/DL algorithms also analyze how students perceive and explore a course and recommend returning to the same learning course if necessary (this type of e-learning is available in MagicBox, https://www.getmagicbox.com/). Hence, data-driven technologies deliver personalized education and improve outcomes for students, helping them reach their full potential.

Online learning applications are becoming more and more popular in higher education around the world. For example, the SHERLOCK application is used to teach Air Force technicians to diagnose electrical system problems in aircraft [120]. Likewise, University of Southern California's Information Sciences Institute aims to train military personnel being sent to international positions to conduct appropriate behavior when dealing with people from different cultural backgrounds by developing more advanced avatar-based training methods. Personalized learning algorithms like Bayesian Knowledge Tracing enable individualized mastery learning and problem sequencing.

The explosion of MOOCs and other models of online education at all levels is the most surprising, including the use of tools such as Wikipedia and Udacity, as well as sophisticated learning management systems that include synchronous as well as asynchronous education and adaptive learning tools. The hugely popular MOOCs, including those created by Coursera and Udacity, use NLP, ML/DL and crowdsourcing techniques to evaluate short answer and essay questions as well as programming assignments. Online education systems that support graduate level professional education and lifelong learning, such as Udacity and Udemy, are also developing rapidly. They are receiving more and more attention as face-to-face communication has become less important in recent years for working professionals who can pursue the change in career at their own pace. There are also AI-powered tools helping to learn and speak different languages such as ELSA Speak. This application uses short dialogues to teach users how to speak and pronounce English words. ML algorithms within the application help to give instant feedback, hence, helping users to make quick progress.

In addition, AI applications help make global classrooms accessible to every single student, regardless of disabilities or language. For example, Presentation Translator is a plug-in for PowerPoint that displays real-time subtitles in any supported language of what the teacher presents. A text-to-speech (TTS) task consists of transforming a string of input characters into a waveform representing the corresponding output speech. For example, text is highlighted, and it will play the audio from that text. In general, TTS applications are well known as assistive aids not only for educational institutions, but also for consumers, businesses and private users. Fundamentally, the general principle of TTS applications remains the same allowing auditory rather than visual consumption of a digital text. Moreover, modern TTS systems base on complex multi-stage processing pipelines, each of which may rely on hand-engineered features and heuristics. Because of this complexity, the development of new TTS systems can be very labor intensive and difficult. So, there are many approaches as there are many models that can be implemented to replace the traditional text-to-speech pipelines such as deep voice neural network and tacotron models based on DNNs [13, 207]. The biggest issue to consider in a deep voice neural network for TTS synthesis is that they require separate training for many different steps of their TTS pipeline. While for the tacotron approach, it can be fully trained from scratch by random initialization, and it does not require phoneme-level alignment, so it can easily scale large amounts of acoustic data with transcripts.

The voice conversion (VC) system transforms the utterance of a source speaker so that it will be perceived as if it was spoken by the specified target speaker. According to the description of the audiologist, in order to differentiate the sound of one speaker from another, it will first look and analyze the pitch of the speaker and second, it will look at the timbre of the speaker [136]. VC itself is a technique to modify a speech waveform which freely converts non/para linguistic information while preserving linguistic information. The development of this technique requires a deep understanding of how to effectively factorize speech acoustics into its individual components, such as linguistic, non-linguistic, and para-linguistic information, using various technologies like speech analysis, speech synthesis, acoustic modeling and ML/DL. In addition, VC has great implementation in a variety of applications, especially in medical aids and education. In education, it can help design hearing aids suitable for students with special hearing problems. In addition, VC may help the student in learning foreign languages, especially in pronunciation exercises [89].

Therefore, in the very near future, AI and DL will have a strong place in all educational experiences. As already noted, DL has begun to prove its effectiveness in a wide range of educational areas, listed below:

- Adaptive learning,
- Automating the evaluation and administrative tasks,
- Predictive analytics,

- Teaching assistants, personalized learning and content analysis,
- Accessible classrooms.

4.4 Healthcare

With the help of ML/DL, clinicians are using AI to forecast the spread of certain diseases and predict which patients are most likely to succumb to it. Preventive care can then be offered with this technology. Hospitals are able to leverage this technology to schedule staff members, set budgets, and increase inventory levels, thereby optimizing the overall healthcare process and minimizing loss. Finally, the shift towards preventive medicine could be implemented and the costs of healthcare per person could be reduced, since expensive health treatments could be prevented through preventive medicine. With patient history data, through AI-powered diagnostic tools and environmental factors, that could potentially influence patient health, doctors will be able to diagnose patient health problems based on factors that could be overlooked.

Healthcare organizations around the world are showing increasing interest in how AI can support better patient care while reducing costs and increasing efficiency. As a result, IT companies are leading the edge of clinical decision support. For example, Google's DeepMind has published the initial results for its DL clinical decision support tool, which can identify more than 50 eye diseases based on the automatic detection of relevant eye features [43]. In the future, virtual assistants equipped with speech/image recognition and DL tools will be able to assist in consultations and full diagnoses. Prescription of drugs could possibly be implemented with the same technologies as well. This will greatly reduce waiting times and allow more effective use of human doctors for more serious and complex diagnoses. Personalized treatment plans designed by DL can be used to greatly improve the effectiveness of therapy by personalizing treatment according to the needs of specific patients.

Object tracking is one of the most important components in a wide range of applications such as surveillance, human-computer interaction and medical imaging [217]. It is built on the merging trackers or mining context information [240]. Object tracking is applied through image sequences and achieved by concentrating only on boundaries, such as apparent contours. This helps reduce computational demands and allows frame rate tracking on non-specialized hardware [217]. While there has been a lot of research on object tracking, there are many factors that affect the performance of a tracking algorithm, including illumination variation, occlusion, and background clutters, and there is no single approach that can handle all scenarios. The appearance model is used in a fixed frame to represent an object with the proper features and verify predictions using object representations. For successive frames, a motion model is applied to foresee the likely state of an object [234]. Object detection works with the adaptive background modeling module, which deals with changing illuminations and does not require constant movement of objects.

Similar to classification, localization can be used for smart cropping or even for regular object extraction, but the problem arises when both of these approaches need to detect and classify multiple objects at the same time. The output of object detection is not fixed in length, as the number of detected objects may vary from image to image due to classification approach.

Fast R-CNNs are similar to R-CNNs, and they use selective search to generate object proposals, but instead of extracting them all independently and using SVM classifiers, they apply CNNs to the complete image and then use regions of interest (RoI) and pooling on the feature map with a final feed forward network for classification and regression. This approach was not only faster, but having the RoI Pooling layer and the fully connected layers allowed the model to be end-to-end differentiable and easier to train. The biggest downside was that the model still relied on selective search, which became a bottleneck when used for inference.

A correlation filter design known as a technique for detecting a known object in a random noisy image. This technique is usually used for object localization, tracking, security and surveillance, object alignment, object detection and classification. Besides that, this technique also allows the computation of an image relative to itself or any number of images, if the computational overhead associated with so many comparisons is feasible on the hardware. There are three different noise models to find the output of peak to side-lobe ratio (PSR) and peak to correlation energy (PCE) value [47]. The noise models include the correlation value found in the absence of noise. Then the correlation value is found in the presence of several types of noise, and the noise is removed from the query image or a given image using various filter design methods. In conclusion, this paper has observed PSR and PCE values under various noise conditions. The noise performance can be improved and will require filters to minimize the noise energy, therefore, improving object recognition performance.

The detection and recognition of organs and tumors in medical images is a prerequisite for medical applications. It is also a crucial step in medical image processing where the goal is to interpret the content of the image. Thus, it combines segmentation or object recognition techniques using prior knowledge of the expected image content. Consequently, it is potentially the most fruitful application area of NN, as it allows to roll several of the preceding stages (preprocessing, segmentation) into one and train it as a single system. It is widely used to analyze images such as MRI results or X-rays. Several CNN approaches surpass the accuracy of human diagnosticians in identifying important features of images [103]. With the spread of the COVID-19 pandemic, it becomes important to identify cases early to prevent the further spread of the virus. In [3], the authors proposed a deep CNN model for detecting the virus using chest radiographs. They increased the depth of the FocusNet based encoder [99] by adding a fourth residual-strided residual block with a single Squeeze-Excitation layer (Fig. 4.11). Compared to early research work, the FocusCovid model achieved a better accuracy of 99.20%. This could be further elaborated to differentiate COVID-19 from other types of pneumonia.

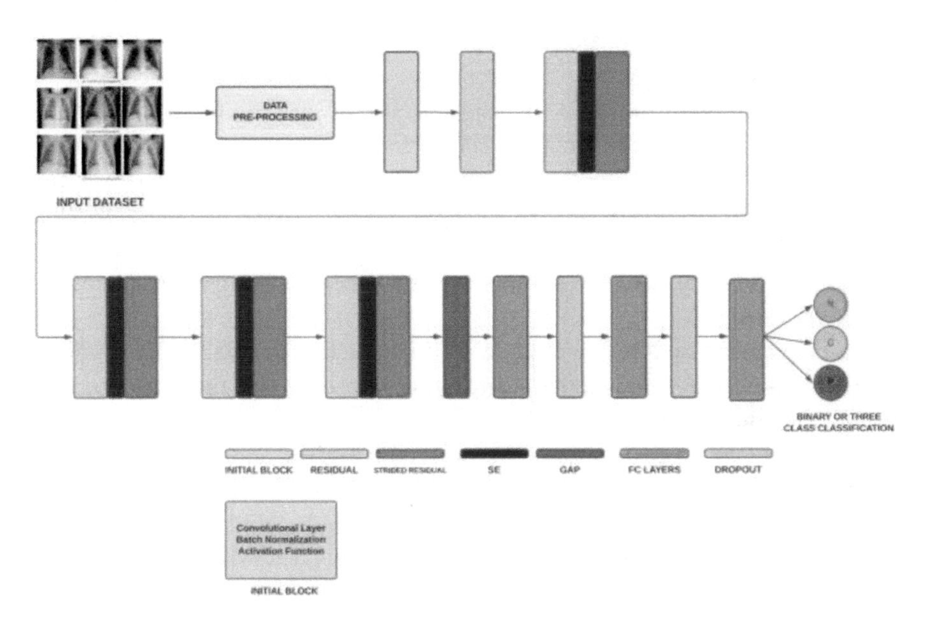

Figure 4.11: FocusCovid architecture [3].

Since dermatology relies heavily on image processing, there is a connection with DL solutions. For example, SkinVision is developed to detect the most common types of skin cancer based on ML algorithms. Once verified, this app also provides some recommendations on the next steps (https://www.skinvision.com/). In recent years, CNN based solutions have been widely used to diagnose and classify skin cancer without preselection of suspicious lesions by dermatologists [52, 73]. These solutions provide high accuracy with a level of competence comparable to that of dermatologists. They are used not only for skin cancer, but are also used to detect and classify different types of cancer [98, 177, 131]. Such techniques are also used to detect and assess pain level based on facial expressions [18]. In [128], researchers detect pain level by analyzing functional near-infrared spectroscopy signals. The results confirmed the importance of a multi-task learning model for personalizing assessment.

Text recognition in real-world images is an open problem that has received significant attention since it has become a critical and crucial component in several computer vision applications such as searching images by textual content, reading labels on businesses in map applications, or assisting visually impaired people. Since text is a pervasive element in many environments and is widely used, solving this problem has a potential for a significant impact [142]. Word detection and recognition is commonly performed by two approaches: a state-of-the-art text detector and a leading Optical Character Recognition (OCR) tool [205]. Factors contributing to the complexity of text recognition include non uniform backgrounds, the need for compensation of perspective effects, short snippets written in different fonts and languages, and proper names that prevent effective use of a dictionary. We can conclude that text recognition receives attention, as it can give many benefits such as assisting visually impaired people, and it has been implemented in Google Street View. Text recognition can produce a good result if there is no complexity such as a non-uniform background or a blurry image.

Early diagnosis and treatment of development delays in children may have a wonderful impact on their emotional and physical health in the future. For example, in [162], the authors have developed a system that detects language disorders at an early age to suggest successful speech therapy. The approach is based on glottal and acoustic features to identify children with specific language impairments using a speech signal (Fig. 4.12). The results indicate that glottal features combined with Mel-frequency cepstral coefficient and openSMILE-based acoustic features provide better classification accuracies. In addition, the feed-forward NN solution outperforms the SVM classifier method when working with combined features, achieving high accuracy of 95%–99%.

Along with helping tools for people with disabilities, there are also DL-enabled chatbots and health assistants that help monitor emotional health, sleep quality, and overall well-being. These tools are changing the way patients interact with the healthcare system by offering 24/7 access to basic health assistance and home-based chronic disease management programs. For example, the Youper application helps users relax with personalized meditations (https://www.youper.ai/) that are achieved through DL solutions. In [159], the authors proposed a DL-based system that helps improve lives of people by checking for common disease symptoms, exercise recommendation, and more. These

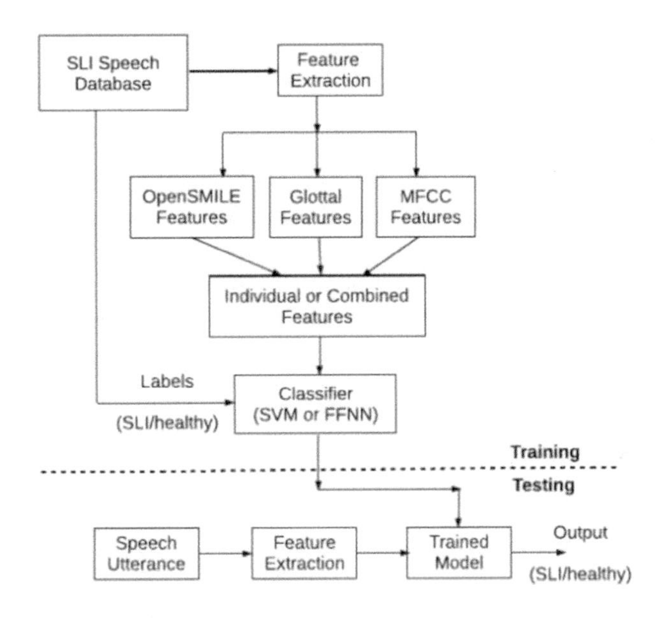

Figure 4.12: The proposed model to detect specific language impairment [162].

types of health assistants are important in shortening the waiting time for a doctor's visit.

In recent years, AI has been widely used for remote patient care based on information received through sensors [152]. These wearable devices allow constant monitoring, and the obtained data is used to suggest a healthy lifestyle. This branch of AI offers the ability to analyze this massive amount of data with high speed and accuracy [55]. AI is also useful for helping take care of the older population. For example, not only personal sensors, but also environmental sensors can identify issues and alert caretakers in case of abnormalities [155].

Precision medicine and drug discovery also rely on the advancement of AI. These tasks require processing an enormous amount of genomic data to identify connections between genes and the physical environment. Hence, DL can also be used here to discover these new patterns, especially when data sets and final results are undefined. It has been shown that by using NNs, adverse drug events can be detected with much greater accuracy and speed [40].

AtomNet is an example of a CNN-based drug discovery system that can predict the binding of small molecules to proteins [203]. It was designed for drug discovery applications by identifying a safe drug candidate from a database. There is a lot of research towards drug discovery based on the advancements of ML/DL. A good example is the creation of six novel inhibitors of the DDR1 gene in less than two months. Generative Tensorial Reinforcement Learning system has shown positive results in mice [232].

Pure medical applications are just a small part of how ML and DL are changing the current functions of the healthcare system. One interesting DL application developed by Google involves the concept of Deep Dreaming, which makes a machine to hallucinate on top of existing database images. This technique is used to simulate biologically plausible hallucinations, allowing users to have more induced dreaming experiences through virtual reality [191]. In addition, such deep dreaming NN models can be used to predict psychosis [100].

Thus, AI is reshaping the healthcare industry as well. It is offering a number of advantages over traditional analytics and clinical decision-making techniques. DL models help identify abnormal patterns in medical images, highlight relationships between symptoms and outcomes, cluster patients, and more. These algorithms are becoming more accurate every day as they interact with more training data. Consequently, DL is steadily finding its way into innovative applications in real-world clinical environments. Below are the healthcare industry areas where AI is already having a major impact.

- Disease prediction and preventive care,
- Medical diagnostics and pain management,
- Health assistants and remote treatment through sensors,
- Dreaming experience,
- Drug discovery.

4.5 Legal and Politics

The legal and political industries are probably the least developed from a technological perspective. They are very slow in introducing innovative technologies and advances in AI. These sectors have unique challenges compared to other industries in programming machines to perform similar to humans. This can be explained by the fact that there is rarely a single mathematical approach to these challenges, and there is still no full confidence due to the fact that AI is a black box. One way forward could be the creation of explainable AI, which will offer complete visibility into why an AI-powered decision is being made. Lawyers and politicians exist because of nuances, disagreements, and unusual situations. Their constantly changing nature and the unique patterns of each individual and unpredictable situation mean that AI can not bring fresh air to these industries. However, many professionals in these sectors are now using AI applications in their day-to-day work without even realizing it. While AI cannot achieve everything that these professionals in their field do, it can still help reduce the time and brainpower for some tasks, even those that require creative thinking and strategy.

Law firms usually spend a lot of time on tasks such as legal research or due diligence, and there is already an opportunity to automate such legal tasks. In addition, giving research recommendations will help lawyers work faster and only apply their knowledge and expertise to higher-level tasks, such as advising clients or negotiating deals. Document review and due diligence usually require the ability to scan a huge amount of documents in search of particular information. DL solutions for categorization and contextualization can help achieve this goal. AI applications can review documents, flag them as relevant so that they can be used to find similar documents. In [75], the authors designed a DNN with additional NLP and word2vec to retrieve the relevant article for a given yes/no legal question to a Japanese exam bar query. They achieve a relational pattern between the legal question and the final article.

Obtaining background information on behalf of their clients is a legal task that takes too much time from experts. Due diligence includes a thorough evaluation of the case and confirmation of the facts to advise clients. AI tools can help professionals more accurately perform due diligence [165]. For example, the ROSS Intelligence tool was designed to behave more like consulting an associate than performing queries on a search engine (https://blog.rossintelligence.com/). Such tools can also be used in politics to detect fake information over the internet.

In the legal industry, AI also assists in the contract drafting process. For example, Evisort is an AI platform designed to extract contract data, identify contract risks, and even provide data analytics for contract generation (https://www.evisort.com/). There are also AI tools specifically designed to review and analyze both contracts in bulk and individual ones. Notable examples are LawGeex, Kira Systems and GLS LegalSifter, which help review contracts faster with fewer errors than humans (https://www.lawgeex.com/, https://kirasystems.com/, https://www.gls.global/gls-legalsifter-ai-powered-legal-document-review).

In legal and politics, AI can be used to analyze data to make predictions about the outcome of legal proceedings or elections. Having access to years of experimental data on various cases and situations, as well as using the computational power of current machines, professionals will be able to make recommendations to clients in the shortest possible time. In [28], the authors have released a data set of legal predictions from the European Court of Human Rights cases. The goal is to predict the outcome of a court case, taking into account textual legal information. They evaluated several neural models (hierarchical attention network, label-wise attention network, BERT and HIER-BERT) on this data. The HIER-BERT model was able to predict whether any human rights article was violated with a precision of 85–90%, depending on the anonymity of the data. To predict which specific article was violated is more difficult, and for HIER-BERT it achieved approximately 65%. Hence, these models can be used by law firms to quickly check if a case is promising or not.

In [122], the authors have proposed a new multichannel attentive NN (MANN) for predicting a legal judgment based not only on the facts of the case, but also on the defendant person. The MANN framework consists of an input layer based on fact description and defendant person, multichannel attentive encoders for capturing informative words and sentences to generate the fact embedding and output predictors for deciding the most relevant law articles. The overall results show that this framework outperforms other solutions in real-world criminal data sets in mainland China.

With the advances of technology, everything has changed: the way we live, interact with each other and even vote. Technology actually plays a huge role in politics, including social media, misinformation and algorithmic bias. In recent years, forecasting of the presidential election outcomes based on the preferences of voters has become very popular. In [241], the support vector regression model outperforms ANN approach, successfully predicting the outcome of the US elections for 2004, 2008 and 2012. For the model, various independent variables

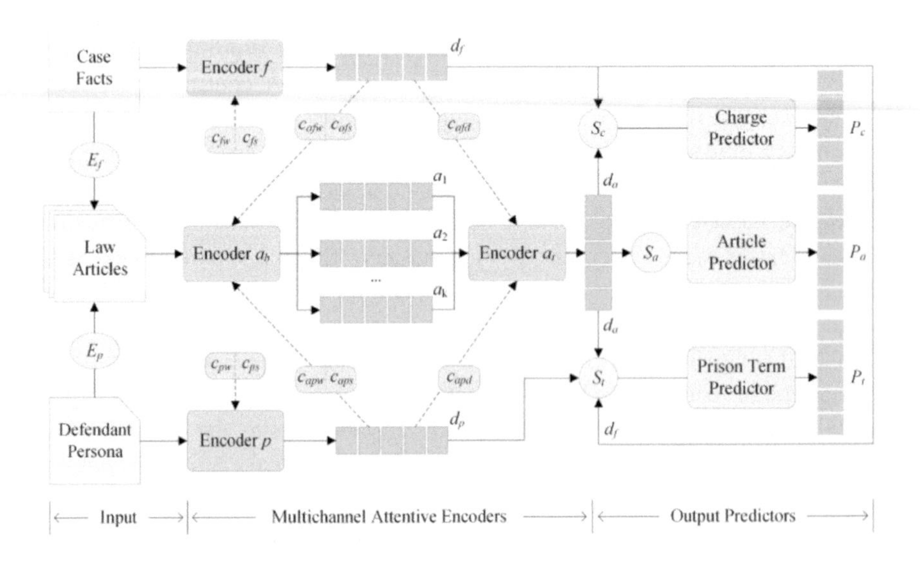

Figure 4.13: The MANN framework [122].

were considered, including personal income, gross domestic product, unemployment rate, changes in the votes of the incumbent party in the last congress election, and the president's job approval. The results confirmed the theory that the presidential election is a referendum on the incumbent president's policies. Another research showed 90% accuracy, finding a correlation between popularity on social media and the final outcome [29]. Deep CNN is used to classify place names by state, and the model is applied to the subscriber "location" in Twitter for each presidential candidate. This model has correctly chosen the president in 45 out of the 50 states.

As with other industries, in these sectors there are also chatbots and virtual advisers based on ML/DL solutions. For example, in China there are legal bots that address small legal issues through WeChat. In 2017, the Hangzhou internet court was established, where citizens use video messaging to communicate with AI-powered judges. Another example is the DoNotPay application, which is a robot lawyer capable of solving all kind of standard legal issues (https://donotpay.com/). Personalized politician virtual agents have been predicted to be the next major thing targeting voters in the upcoming presidential elections. The ability to interact with at least digital representation of the president's personality in the form of a virtual agent will significantly affect voters. As a first step, the world's first AI-powered virtual politician was created in New Zealand. This chatbot, known as Sam, was designed for the 2020 general election to analyze citizens' opinions and to address queries regarding local issues (http://www.politiciansam.nz/).

The legal profession, along with the politics, continue to be the least inclined to incorporate AI into their business models. However, there are already areas in those sectors that AI and DL are already transforming, including:

- Legal research and due diligence,
- Contract review and risk analysis,
- Legal judgment and election predictions,
- Virtual advisers.

4.6 Military and Security

AI is deployed in almost all applications in the military sector. It is obvious that AI, together with innovative automatic systems, will become an inseparable element of future armed conflicts. ML and DL algorithms are intensively used not only in information warfare, but also in land, sea and air warfare.

Military services around the world are trying to embed AI into autonomous vehicles, weapons and defense systems. For example, Israel Aerospace Industries has designed a range of autonomous military vehicles. The RoBattle is an example of an autonomous tank capable of eliminating many obstacles. It can be used not only for information gathering and force protection, but also for advanced attacks. The company also offers an autonomous robotic solution, RobDozer, which can be used to overcome road challenges, including removing large objects and excavating. These vehicles help carry out key tasks without risking human lives.

Computer vision is a key part for controlling autonomous weapons such as combat drones, killer robots, etc. Such weapons usually have fully automated actions without any human intervention. They also use algorithms that provide greater accuracy in object recognition, as accurate image processing and fast target detection are critical to military systems. For example, the Defense Advanced Research Projects Agency's (DARPA) Target Recognition and Adaption in Contested Environments program uses advanced ML techniques to automatically locate and identify targets based on Synthetic-Aperture Radar (SAR) images. These algorithms also combine GPS data with field data to refine the target as much as possible and avoid collateral damage. Therefore, such target detection algorithms need to address several challenges, including recognition speed and poor edge detection [106]. In [105], the authors proposed the EdgeBoxes algorithm for object detection and CNN for classification. CNN models have been proven to be invariant to minor rotations and shifts in the target object (Fig. 4.14).

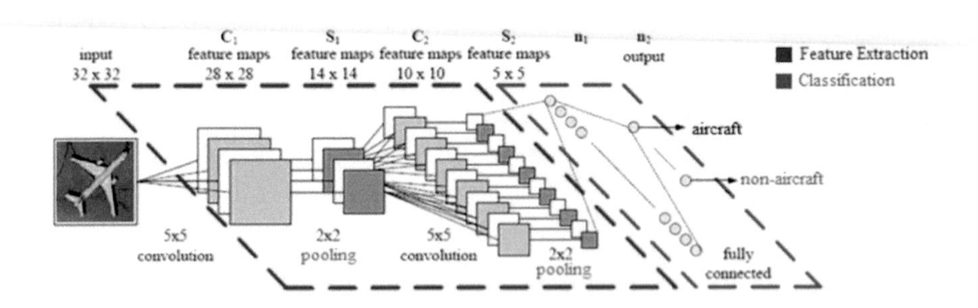

Figure 4.14: CNN based model for target detection [105].

The detection of underwater mines in anti-submarine operations is another important topic in the defense sector. The purpose of this operation is to locate suspicious objects on the seafloor and then classify each of them as a mine. This is currently achievable with an autonomous underwater vehicle (AUV) that has sonar sensors capable of producing very high resolution images. The 10-layer deep CNN can then be used to classify these images [211]. The authors have shown that the use of DNN is justified because they are capable of automatically learning valuable differences between similar classes of objects. This is especially important for the classification of underwater images. Similar techniques are also applied for landmine detection [119]. Another approach, which uses thousands of images of two different types of anti-tank mines, showing them covered, partially covered, and upside down from various angles and in various lighting conditions, is presented in the work [2]. This tool achieves a 99.6% success rate when detecting non-hidden landmines.

Many cities have already begun to deploy AI technologies for public safety and security due to its high efficiency and accuracy. According to a research conducted by Stanford University, a typical North American city will rely heavily upon them by 2030. Technologies applied in these fields include cameras for surveillance that can detect anomalies pointing to a possible crime, drones and predictive policing applications. AI may enable policing to become more targeted and only used when absolutely necessary. This will significantly reduce the costs of governments spent on security and safety. When deployed carefully, AI may also help remove some of the biases inherent in human decision-making.

Facial recognition applications are widely used in various sectors. As we discussed earlier, it is progressively being used in the banking industry. However, it is important to note that such applications play a key role in the fight against terrorism. The Active Face Recognition (AcFR) system actively recognizes faces by mimicking human behavior where it employs a CNN, and acts consistently with human behaviors in common face recognition in every possible situation. This system contains two main components, namely face recognition module and a behavior controller module where face recognition module employs VGG-Face CNN evaluates the input image and provides enough information to make decisions [140]. It also makes the visual processing more biologically reasonable, which can be used for future applications of AcFR. The controller module uses this information retrieved from the behavior controller to drive its follow-up behavior when reassessing with the subject. It is also designed to perform facial recognition in a view-driven sequential manner.

Corsight AI company has proposed a facial recognition tool that can identify individuals even when part of their face is covered. Facial recognition tools are also associated with DL for better risk detection. Algorithms are trained to understand normal behavior and send alerts in case of abnormal behavior to ensure public safety. In [193], the authors suggested the CNN-LSTM method for extracting important features from video clips and predicting abnormal behavior in both two-people and crowd-based interactions (Fig. 4.15).

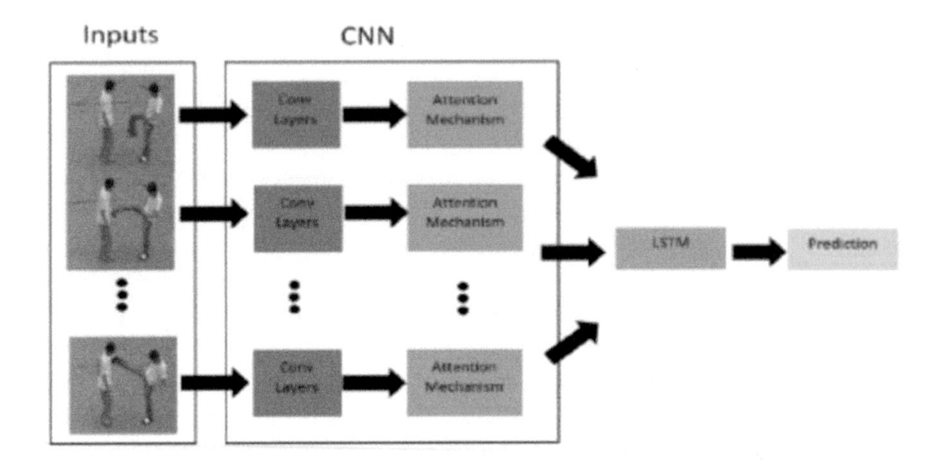

Figure 4.15: CNN-LSTM model for classifying behavior [193].

The New York City Police Department's CompStat was the first tool pointing toward predictive policing in the United States, and many police departments in the country now use it. ML significantly enhances the ability to predict where and when crimes are more likely to happen and who is likely to commit them [104]. Widely deployed AI prediction tools have the potential to eliminate or reduce human bias rather than reinforcing it, given that research and resources are being directed toward ensuring this effect.

Simulation and training for police and army forces is a multidisciplinary field that constructs computerized models that acquaint personnel with the combat systems deployed during various operations. Currently, countries around the world are investing more and more in such applications. In particular, military training software is a frequently used type of simulation application because it is the best way to simulate real-life situations in the air, navy, marine, etc. For example, the Advanced Training Management System (ATMS) has become one of the best military simulation and training applications designed for the USA. In addition, the army forces have been employing virtual reality software to train their soldiers for years. ML algorithms are used to develop autonomous agents in such software [166], which is critically important for mimicking real-world combat as closely as possible.

AI techniques can also be used to conduct intelligent simulations to train law enforcement officers so that collaboration with police forces from different countries to join forces to fight international criminal organizations and terrorists from different countries could be conducted. This will help reduce the amount of effort to train international groups of law enforcement personnel to work as a team. Through the Horizon 2020 program, the European Union is currently supporting attempts in projects such as LawTrain. Simulation of actual investigations by providing tools that support such collaborations could be the next step in terms of further development. There are already some tools that exist for scanning Twitter and other feeds to look for certain types of events and their security implications[91, 216]. Besides that, law enforcement agencies are putting more and more efforts to identify plans for disruptive events on social media, as well as to monitor the activity of large gatherings of people in order to analyze security. Crowd simulations have been significantly used to determine how crowds can be controlled.

Several US agencies such as the US Transportation Security Administration (TSA), the Coast Guard, and many other security agencies that currently rely on AI are likely to increase their reliance to enable significant improvements. AI techniques such as vision, speech analysis and gait analysis can be used to aid interviewers detect possible deception and criminal behavior. For example, TSA currently has an ambitious project to redesign airport security nationwide called Dynamic Aviation Risk Management Solution (DARMS). DARMS system is designed to improve the level of efficiency of airport security through the use of personal information and customize security based on a person's profile. The potential future advancement of this project is a tunnel that checks people's security while they pass through it.

Nowadays, logistics, especially in the military domain, includes processing large amounts of data and moving goods or services to a designated location at an agreed time, supporting combat units and many others. The organization of an efficient distribution and supply chain is very important not only in times of war, but also in peacetime. The ML and DL algorithms can be used extensively to predict the availability and reordering of military logistics [5]. In the field of military logistics, the organization of medical care is another crucial topic. In recent years, AI has also been used to monitor and diagnose wounded soldiers in order to provide necessary assistance. The Automated Ruggedized Combat Casualty Care (ARC3) is a good example of a care system that can be used when evacuating a wounded soldier is not possible.

Cybersecurity, as well as issues like spam, are widely shared concerns, and ML/DL is having a huge impact on them. AI tools may help police manage crime scenes or search and rescue events by helping commanders prioritize tasks and allocate resources when tools are ready to automate such activities. Nowadays, hacker attacks are becoming very common not only for private companies within the country, but also for cross countries. The truth is, in current warfare operations, a large percentage of military actions have been transferred to cyberspace. And to ensure the success of a military mission, the security of combat data is playing a critical role. However, in recent years, professional hackers have gained access to the military systems of other countries and damaged them. The military sector is vulnerable to cyberattacks that can result in the loss of military information. This is why the military authorities pay attention to cybersecurity, equipped with AI solutions that can automatically protect data, network and information from any unauthorized access. Improvements in DL in general and transfer learning are the best candidates in enabling this to be realized, and it is currently widely used in cybersecurity domain [219]. In addition, AI-enabled web security systems can record the pattern of cyberattacks and develop counter-attack tools to tackle them.

Current industry standard malware detection systems (often deployed as antivirus technology) deploy signature-based methods of detection which lack the capacity to keep pace with the massive amount of malware samples released daily, as well as their variability [225]. Behaviour-based techniques have proven to be a feasible alternative as they operate on dynamic feature analysis using instruction sequences, computation trace logic, and system (or API) call sequences, however their current usage suffers from a high false-positive rates [225, 215, 220]. Generally, the accuracy of malware detection and classification depends on obtaining sufficient context information and extracting meaningful abstraction of behaviors. DL models are well suited for analyzing longer sequences of system calls and making better decisions through higher level information extraction and semantic knowledge learning. However, the computational time for training these models to estimate the probability of detection in real-time application is intensive.

A novel Android malware detection system which shows the potential of the CNN was introduced in [134]. The system has been shown to perform reliably when compared to other modern technologies and tested against four different sets of malware data. In addition, it is reported that the system is capable of simultaneously learning to perform feature extraction and malware classification when given only the raw opcode sequences of labeled samples. This attribute of the system has not only eliminated the need for manually-crafted malware features, but also proved to be more computationally efficient than existing n-gram based malware classification systems.

With the advancement of technology, mobile phones and computers have become the main part of our daily life. DL and ML methods have been widely used to protect personal data on these devices. For example, Hopfield NNs can be used to detect unusual usage patterns to automatically lock a mobile phone [86]. Cloud computing has also started to be used in recent years, even by large companies. Therefore, it is critical to strengthen security measures to protect data in the cloud infrastructure. DL also helps here predict these issues in advance based on ANNs [49].

Today, AI technologies are not mature enough to radically change the nature of warfare. There are several key challenges that could potentially slow down or limit the full integration of AI in military applications. These include limited training data, vulnerability, insufficient transparency and, of course, ethics. However, in the field of security, DL solutions were used massively. Nowadays, both the military and security sectors are evolving rapidly, and several countries have relaunched the arms race and security systems that rely partially or completely on AI. As already noted, DL has begun to prove its effectiveness in nearly all military and security areas, listed below:

- Warfare systems,
- Target detection,
- Simulation and training,
- Logistics,
- Public safety,
- Cybersecurity,
- Data protection.

4.7 Service and Marketing

AI applications are very popular in marketing and customer service as well. With the vast amount of information and advertising on the internet nowadays, the only way marketers can catch the attention of smart consumers is through personalized advertisement. Market analysts can deliver highly targeted and personalized advertisements based on AI-powered pattern recognition, behavioral analysis, and other tools. They can also re-target the audience at the right time to achieve better results with less annoyance. It is important to note that the combined analysis of texts and images allow more accurately predict the interests of users. In [83], the authors proposed a hybrid model in which CNN models were used to classify images and RNN models were used to classify textual data (Fig. 4.16). Outperforming text-only or image-only classification models, this model can be useful for marketers to make interest-based, rank-order or real-time recommendations. DL can provide real-time personalization based on user posts over the internet and can be used to optimize marketing campaigns according to the needs of the local market. Hence, predictive analytics are also used

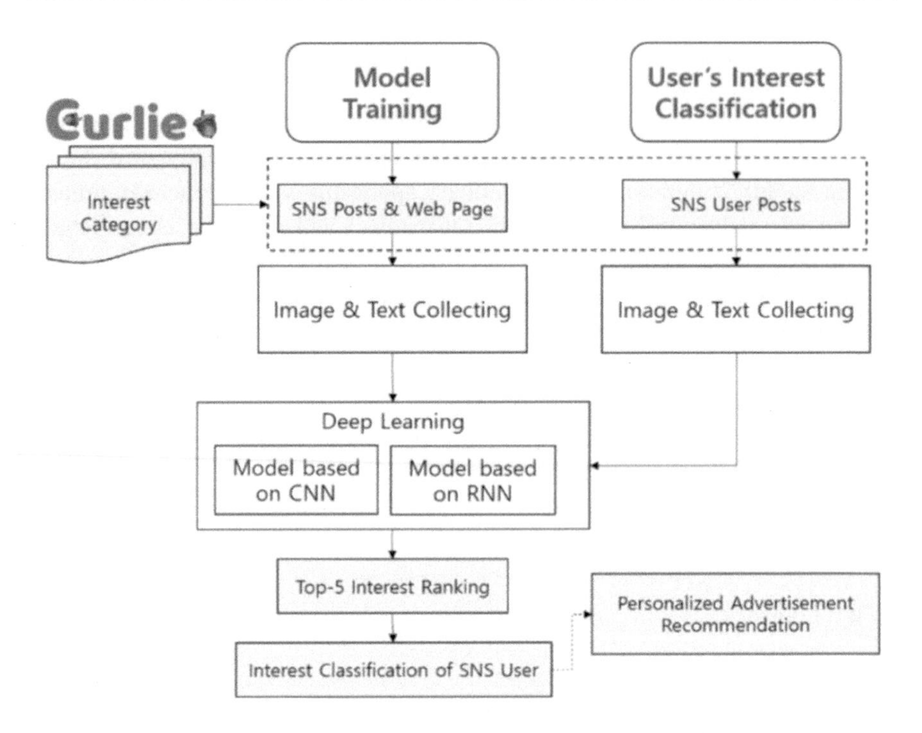

Figure 4.16: Personalized ads classification system [83].

in marketing to engage customer service to handle their target's requests. The theories and mathematical formulas for predictive analytics in marketing have been around for several years, but DL solutions have made them more accessible nowadays.

By optimizing pricing and personalizing promotions to a targeted audience, Geospatial modeling has successfully improved micro market attractiveness by increasing sales. Dynamic pricing and personalization have also increased online sales significantly. As natural language develops, in-store virtual assistants could utilize facial, voice and speech recognition to analyze customer shopping history, make suggestions and communicate in a conversational way using natural language generation. Stitch Fix, an online personal shopper service, utilizes an algorithm to understand clients through the images clients display on Pinterest. On the other hand, using DL, Pinterest is converting the repository of images into potential sales by allowing users to zoom in on specific items within an image, some of which could be purchased directly from the merchants.

Cloud service is subsequently going to enable more rapid release of new software for home robot programs and more sharing of data sets collected from many different points. Great advances in DL will enhance the way people interact in their homes. Low-cost 3D sensors have improved the work on 3D perception algorithms by thousands of researchers worldwide. This subsequently increases the speed of development and adoption of home and service robots. Over the past few years, low cost and safe robot arms have been introduced to hundreds of research laboratories around the world, stimulating the creation of a new class of research on manipulation that will eventually be applicable in the home in the near future. Dozens of startups around the world are developing AI-based robots for services, currently concentrating mainly on social interaction [196, 151].

As such, the customer service industry is no exception when it comes to using DL methods. Customers have direct access to product information, and businesses have access to customer data, which they later use to reach out potential clients. Therefore, it is very important for businesses to offer seamless customer service experiences. This is where virtual assistants and chatbots help by handling multiple queries at once, which is an advantage for businesses with busy call centers. In case the question is too complex for a chatbot, only then can it be transferred to a live representative. For example, H&M, a clothing retailer, uses the Kik messaging application to help customers shop entirely through a conversation with a bot. Sometimes it is even hard to tell if you are talking to a human or bot. However, if the question has been answered quickly and correctly, then it really does not matter.

On the technical side, virtual assistants are a fundamental evolution of a Visual Question Answering (VQA) system based on NLP and computer vision [214]. Computer vision is used to understand the image, and NLP is implemented for understanding the question and answer. Moreover, this task typically shows the real-world image to a computer and asks a random question. The answer forms include a number, a yes/no answer, a choice out of the several possible answers, and a fill in the blank. The DL methods have shown state-of-the-art results in VQA systems [71]. The subsets of NLP including sentiment analysis and text categorization will be thoroughly discussed in Section 4.8.

AI takes data and uses it to identify the best representative available to address customer needs. These approaches can be applied to both call and message routing systems. RNN and SVM can be used to classify spoken languages in help desks [60]. Several studies show that using large amounts of business data along with the LightGBM algorithm can improve the prediction of the problem for which a customer is calling [94].

There are many popular DL-based virtual assistants nowadays, ranging from Alexa to Google Assistant. These tools learn and collect information about the user's preferences with every interaction. They are designed to provide personalized assistance that can translate speech to text, make notes, auto-respond to your specific calls, create and send appropriate emails, etc. Such assistance is also used in various industries (https://www.makerobos.com/). In real estate, AI-powered bots help agents find the perfect match for people looking to rent, sell or buy properties. The travel industry derives significant benefits from AI-enabled chatbots to process bookings and inquiries of passengers. Collecting user preferences can help recommend different options in the travel domain. For example, Oscar is an Air New Zealand's chatbot that answers customer questions about 67% of the time. Robots of the more physical variety are already present in airports and hotel lobbies, assisting travelers. For instance, Connie is a robot concierge that can interact with hotel guests using speech recognition to respond to their queries. It uses ML algorithms to learn from each interaction.

In [12], the authors have proposed another travel engine that interacts with users via Amazon Alexa and do the appropriate recommendations using the Restricted Boltzmann Machine (Fig. 4.17). Another example of a personal virtual assistant is DataBot, which answers questions related to topics that matter to users (https://www.databot-app.com/). It can also provide images and presentations upon your request. It speaks different languages and retrieves information from Wikipedia, Google search, etc.

User experience enhancement with the help of AI technologies is endless. The combination of DL and computer vision can help store owners compete with online retailers by eliminating the checkout process versus the one clicks checkout process currently adopted by e-commerce. Amazon Go in Seattle has experimented with a futuristic user experience, using computer vision to identify customers as they enter the shop and link them with products taken from shelves. It automatically deducts the cost of items purchased by customers from their Amazon account. This enables customers to enjoy the experience of purchasing items physically without the hassle of queuing up at the cashier for payment as well as having to look for items needed one by one along the aisles.

Insurance companies now see DL as a promising tool for handling claims and customer service. As discussed earlier, ML/DL algorithms can easily recognize patterns in fraud, which can help insurance companies differentiate between false and legitimate claims. For example, Tractable suggested DL solutions for automobile insurance (https://tractable.ai/). It uses images of damaged cars to estimate future repair costs, which reduces time spent on visiting the shop. Similarly, PwC uses DL for image analysis to estimate repair costs after a customer has filed a claim (https://www.pwc.com/).

The two main schools of information retrieval (IR) are generative and discriminative models. While generative models successfully model features, they suffer from the difficulty in leveraging relevancy signals from other channels like links and clicks. On the other hand, discriminative models lack a principled way of obtaining useful features or gathering helpful signals from massive unlabeled

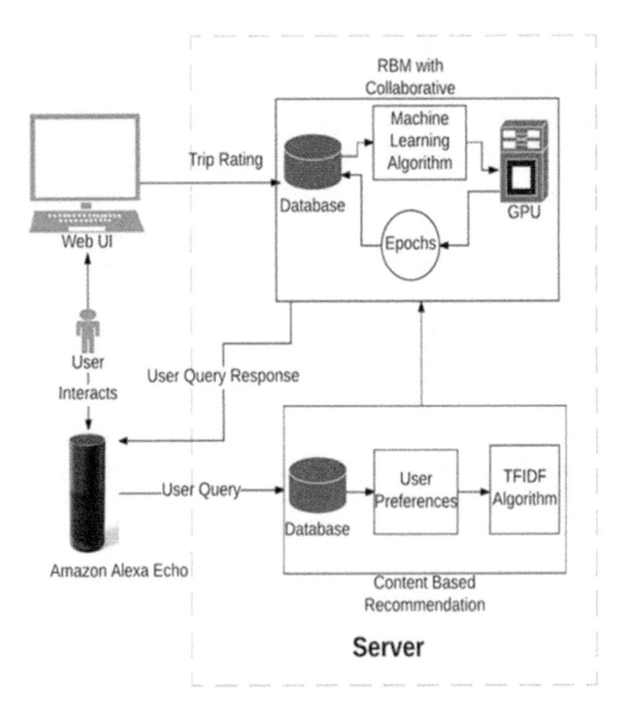

Figure 4.17: Recommendation system architecture [12].

data. Inspired by the Generative Adversarial Nets (GANs) [65], the Information Retrieval Generative Adversarial Networks (IRGAN) have combined these two models. They focus on predicting relevant documents given a query in generative retrieval modeling and predicting relevancy given a query-document pair in discriminative retrieval modeling. It then uses the minimax game adversarial training on the predicted parameters so that a unified optimized framework is produced. This model takes advantage of characteristics of both models, where the generative model acts as an attacker to the current discrimination model, generating difficult examples for the discriminative model in an adversarial way by minimizing its discrimination objective. Hence, this model differs from existing game-theoretic information retrieval methods [129, 228]. Overall, IRGAN provided a more flexible and principled training environment where extensive experiments were conducted on four real-world data sets in three typical IR tasks: web search, item recommendation, and question answering.

The core of personalized service is to estimate the likelihood that a user will purchase an item based on the characteristics of the user, context, and item. Hence, for recommendation systems, predicting user responses is a critical topic and can be seen as a binary classification problem. Predicting user response over multi-field categorical data in a scenario of click prediction has been studied in [157]. Product based NN has been proposed that, compared to previous research works, is not restricted to inferior feature extractors or weak classifiers. In recent years, DL algorithms have shown their effectiveness when applied to the research of recommendation systems [231]. Many travel and e-commerce web sites collect reviews of properties or shopping items. These reviews are later used to extract user preferences and make recommendations accordingly. While very good at extracting complex features, CNNs are widely used to model user preferences and item properties [178].

Human resources provide services to companies and businesses to organize the workforce. AI-driven systems can be used to examine applications based on specific parameters, screen candidate profiles, and match candidates to employers. It has been widely used in the recruitment process since 2018 [200]. For example, IKEA and Amazon have started using recruitment systems like Mya or HireVue to help improve their talent-hiring capabilities. In [188], a CNN-based Asynchronous Video Interview (AVI) platform has been proposed to predict job candidate communication skills and big five personality traits. Figure 4.18 shows a CNN model consisting of four convolutional layers, three pooling layers, ten mixed layers, a fully connected layer and a softmax layer as the output. ReLU was used to fix the vanishing gradient problem. Using this model, communication skills and three personality traits (openness, agreeableness and neuroticism) were predicted successfully, but two other traits (conscientiousness and extraversion) were unsuccessful based on the candidates' facial expressions.

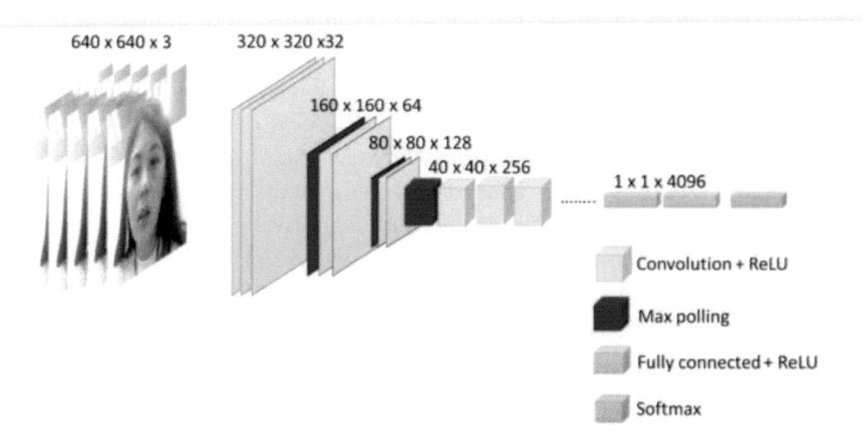

Figure 4.18: CNN prediction model [188].

Therefore, DNNs have tremendous potential to dramatically improve customer service and marketing analysis. AI-enabled tools help individuals and companies solve issues faster and more accurately. Below are the main domains currently affected by DNN solutions:

- Recommendation systems,
- Virtual assistants and online customer services,
- Smart recruitment,
- Intelligent call or message routing systems.

4.8 Social Media and Entertainment

Since the explosive growth of the Internet over the past fifteen years, few people nowadays can imagine their daily life without it. The Internet has established itself as a platform for user generated content as a viable source of information and entertainment. Trusted applications now exist for sharing user-generated content through various forms of information such as blogs, tweets, videos, etc. To operate at the scale of the Internet, these applications must rely on techniques developed in the field of NLP, information retrieval, image processing, crowdsourcing, and ML/DL. Traditional sources of entertainment are also using AI to keep up with the times. Professional sport is now subjected to intensive quantitative analysis, as exemplified in the book and movie Moneyball. On-field signals can be controlled with sophisticated sensors and cameras instead of relying on aggregate performance statistics. Software applications were created for composing music and recognizing soundtracks, along with techniques from computer vision and NLP that were used to create stage performances. The user can even exercise his or her creativity on platforms that automatically generate 3D scenes from natural language text such as WordsEye. Consequently, AI is changing the present and future of social media and entertainment domains, and this is essential to meet the needs of Gen Z and Millennials.

Various brands have used social media listening to get real-time feedback on audience reaction to changes. This helps marketers improve campaign results and crisis management. For example, NetBase uses ML algorithms to monitor millions of social media discussions for better results in their feedback (https://netbasequid.com/). Social media giants like Facebook, Pinterest, Snapchat and others are also using AI-based social media listening techniques to provide more personalized products and services to their users. They use advanced ML/DL algorithms to do everything from recommending content to targeting users with advertising. For example, Facebook's Deep Text tool uses ML algorithms to learn context by monitoring comments and posts of users [213]. This tool can also recognize posts with suicidal content. Approximately 1 suicide occurs in the world every 30 seconds, so the company can prevent such cases by sending ads to these users with suicide prevention materials. In [145], the authors constructed two DNN-based models to predict suicide risk from Facebook posts. The deep Contextualized Word Embeddings (CWE) algorithm is used to extract representations of texts that are input to NN models. The single task model predicts suicide risk directly from a user's Facebook posts. The multi task model considers personality traits, psychosocial risks and psychiatric disorders as contributing factors. The results confirmed that these risk factors help improve the accuracy of predictions.

With countless pieces of content being created and analyzed every second, classifying these items for later search becomes an important task. These types of classification are used in a variety of domains, from e-commerce to social media. For instance, Netflix is based on ML algorithms to improve the categorization of movies in a way that makes searching easier using category names rather than titles. This is a hard task due to the huge number of existing descriptors, objects and scenes that need to be labeled.

AI-based image classification tools can help analyze the content of video frames and add the appropriate tags. For instance, Google Photos automatically tags all photos that can be easily searched. FaceApp is another example that helps convert a real photo by changing human faces in photographs. This app uses NNs to achieve these transformations. DL algorithms are used to describe every existing item in the photo. In [97], the authors present the DNN method for identifying areas of interest in an image and writing a description for each region (Fig. 4.19). The suggested DNN architecture assumes alignment between the image region and the appropriate segments of sentences. The multimodal RNN architecture is used to generate descriptions of regions, which outperformed retrieval-based models. Such applications can help people with visual disabilities rely on textual explanations of the image.

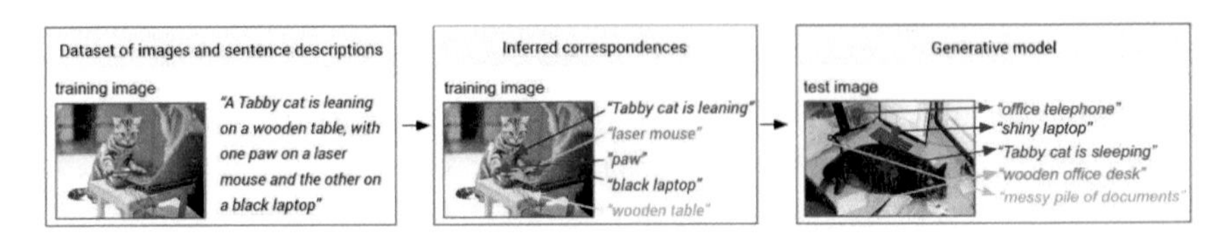

Figure 4.19: DNN method to generate novel descriptions [97].

3D pose estimation is the problem of determining the transformation of an object in a 2D image that gives a 3D object. The need for 3D pose estimation arises from the limitations of feature-based pose estimation. The latest research in pose estimation describes a realistic 3D hand model that represents the 21 various parts of the hand and the use of this hand modeler in the American Sign Language (ASL) digit recognition application [101]. They also describe an SVM-based recognition module for ten ASL digits that attains a recognition rate of 99.9% on live-depth images in real-time using Kinect. With the release of the Kinect, a depth sensor that can work in absolute darkness has made hand pose and gesture recognition easier. The entire system handles the automatic generation and labeling of 2000 synthetic training images by randomizing each skeleton parameter so that it can form large datasets by interpolating poses and perturbing joints without violating skeleton constraints. Multiple decision trees are then trained on separate datasets to perform per-pixel classification, and each pixel is assigned to a part of the hand. Finally, the posterior probabilities of each pixel are combined to estimate the 3D joint coordinates for the hand skeleton. This app can then connect to Kinect and perform skeleton fitting in real-time at 30 fps without experiencing frame drops. It largely depends on the efficiency of the individual randomized decision trees (RDT) for better generalization [23]. By feeding manually designed hand poses corresponding to ASL digits to the random decision forests (RDF), the entire system learned how to correctly classify the hand parts for real depth images, which, in turn, helped researchers collect real data labeled by the RDFs for further pose classification tasks. The system achieved near perfect classification results by mapping each occluded joint to its nearest visible neighbor in the skeleton and optimized for multicore systems also capable of running on high-end notebook PCs without experiencing frame drops. Such an application can be used to improve visual performance skills on institutionalized adults with wheelchairs [35].

When talking about video or image applications, it is important to mention compression methods, which are very important when transferring data. Google researchers have introduced a type of RNN that provides units to save activations and process sequences, named Residual GRU, whereby it combines existing GRUs with residual connections to attain remarkable image quality gains at a given compression rate. In lieu of using Discrete Cosine Transform (DCT), two sets of NNs can be trained, namely the encoder, which creates codes from the image, and the decoder stands to create the image from the codes. This system works by iteratively processing the original image by the encoder and decoder using the Residual GRU layers, where any additional information can be passed from one iteration to another for better-quality reconstruction. The second approach to image compression is known as lossy image compression [195], which is a technique used to reduce the size of data for processing, storing, and transmitting content. It is usually used for data such as image, video, or audio. Autoencoders are unsupervised learning algorithms that apply backpropagation, setting target values to be equal to the inputs. Compressive autoencoders (CAE) are made up of three components: an encoder, a decoder, and a probabilistic model. The goal is to optimize the trade-off between using few bits and little distortion. Researchers achieve better performance than JPEG 2000 in terms of Structural Similarity Index (SSIM) and Mean Opinion Score (MOS). This performance was achieved using an efficient convolutional architecture combined with simple rounding-based quantization and a simple entropy coding scheme.

Zooming into videos or images beyond the actual resolution is only made realistic using DL algorithms. For instance, Google Brain researchers use DNN to predict a person's face from low resolution facial images. The pixel recursive super resolution method enhances the resolution of the image, revealing important features that are enough for image reconstruction or personality identification [38]. The proposed model is capable of increasing the magnification ratio based on complex variations of objects, illumination, and viewpoints. Moreover, sampling multiple times can predict different high resolution images (Fig. 4.20). The model is based on 18–30 series of ResNet blocks that process low resolution images and 20 gated PixelCNN blocks with 32 channels at each layer. There are also applications that allow to completely eliminate the background from the video. For instance, the Unscreen application makes it easy to replace your video backgrounds with new static or even video wallpapers.

Social media companies are also using ML/DL solutions to build automatic translation systems that allow users from different countries to translate the posts

Figure 4.20: Different high resolution images from various samplings [38].

and articles. DL allows us to go further in this direction and makes universal human communication possible. For instance, the Google Translate app translates images with text into your language in real time. The mobile phone camera scans the object and uses the DNN and optical character recognition technologies to convert and then translate it into the preferred language.

Language complexities such as syntax, tonal nuances, or even sarcasm, make it difficult to develop appropriate responses, even with AI-enabled tools. Since their birth, humans develop various social interaction skills that help to give appropriate responses and distinguish expressions for each scenario. NLP with DL methods tries to achieve this feature in order to differentiate linguistic nuances. Text classification, answering questions, twitter analysis or sentiment analysis are all subsets of NLP where DL is currently gaining momentum. The development of techniques for document-level sentiment analysis is one of the significant components of this area. Recently, people have started to express their opinions on the web in a way that has increased the need to analyze opinionated online content for real-world applications.

Most complex neural models in NLP offer little transparency for their inner workings. For certain applications, like prediction used to drive critical decisions, it is important to understand the underlying basis for the decision. Ideal complex neural models should yield improved performance as well as offer interpretable rationales for their predictions. Current approaches incorporate rationale generation as an integral part of the learning problem. These approaches restrict models to extractive rationales by limiting the rationales to be subsets of words from the input text that are (i) short and coherent, (ii) must alone suffice for prediction as a substitute of the original text. The rationale generation must be learned in an unsupervised manner, hence a model with rationales is trained on the same data as the original neural models. The model consists of two modular components, namely a generator and an encoder. The generator specifies the distribution over possible rationales, and the encoder maps such text to task-specific target values. These two components are jointly trained to maximize a cost function that favors short and concise rationales to achieve accurate predictions based on rationales alone. To minimize the ambiguity of what counts as rationales, as well as the difficulty in evaluation of rationale selections, the model focuses on the following domains: (i) concerns with multi-aspect sentiment analysis, (ii) concerns with the problem of retrieving related questions. The model obtained high performance in both tasks, and, for the sentiment prediction task, the model achieved an accuracy of 96%, a significantly higher accuracy compared to the bigram SVM and neural attention baseline.

The model demonstrated that the encoder-generator framework, when trained in an end-to-end manner, gives rise to quality rationales in absence of any explicit rationale annotations. The approach was evaluated on multi-aspect sentiment analysis versus a manually annotated test case, and it significantly outperformed the attention-based baseline.

Questions are usually asked to access knowledge of others or direct one's own information-seeking behavior. Advanced methods teach machines to ask questions due to a few motivations:

- Posing appropriate questions is crucial when collecting information in intelligent systems,
- Learning to ask questions may improve the ability to answer them,
- Answering questions in most existing QA datasets is extractive, while question asking is comparatively abstractive.

The most popular methods today utilize the sequence-to-sequence approach [190] to generate questions that are conditioned on document and answer. The document and the answer are first encoded, and then the question words are sequentially outputted with a decoder that is conditioned on the document and answers encodings. The standard encoder-decoder approach is then augmented with several modifications geared towards the question-generation task. On top of maximum likelihood for predicting questions from (document, answer) tuples during training, policy gradient optimization is used to maximize auxiliary rewards such as language-model-based score for fluency and performance of a pre-trained question-answering model on generated questions. The model not only showed that the policy gradient qualitatively increased the rewards earned by generated questions during testing, but also provided examples to illustrate the qualitative effects of different training schemes. Modern question generation models are recurrent neural models that generate natural-language questions conditioned on text and predefined answers. The model was trained using a combination of maximum likelihood and policy gradient optimization and demonstrated, both qualitatively and quantitatively, how reward combinations affect generative outputs.

Current approaches in text categorization are defined with the associations between word clusters and document clusters for cross-domain learning by proposing a matrix tri-factorization-based classification framework (MTrick) that performs well even when the difficulty degree of transfer learning is great. To capture the features at a conceptual level for classification in MTrick, the associations between word clusters and document clusters remain the same in both the source and target domains. Finally, an iterative algorithm was developed to solve the joint optimization for two-matrix tri-factorizations on the source and target data problem and theoretically prove its convergence. Several tests have been conducted to compare the performance of MTrick with some baseline classification methods, including the supervised algorithms of logistic regression (LR) [84], LibSVM [30], SVM [22], and cross-domain methods of Co-Clustering-based Classification (CoCC) [39], and Local Weighted Wnsemble (LWE) [59]. All of these results show once again that MTrick is an effective approach for cross-domain learning and has a stronger ability to transfer knowledge.

These categorization techniques are used to filter all bad news from users' news feed and customize news according to readers. On the other hand, it can help detect fraud news, which is an important asset nowadays when the Internet has become the primary source of all genuine and fake information [184]. However, this is a difficult task as no one can precisely decide whether the news is neutral or biased. Twitter is also based on DL methods to recommend tweets in user timelines and fight against tweets containing racism, hate speech, or other inappropriate content. It uses IBM Watson [81] and NLP to eliminate abusive messages. IBM Watson is capable of understanding the meaning of messages and different visuals, thus detecting inappropriate content in seconds.

Tasks that have a high probability to be creative are likely the ones that have the small risk of displacement by AI-enabled tools, as they are more resistant to automation. However, there are already entertainment tools that help musicians to generate songs. For instance, Soundraw is an AI-powered tool that creates songs based on existing phrases (https://soundraw.io/). MuseNet is a DNN solution that can generate composition with 10 various instruments and styles by discovering patterns of rhythm and harmony and learning on its own. Another example is a lyric generation tool that offers different lyrics depending on the mood and topic of the users (https://theselyricsdonotexist.com/).

Audio synthesis is the generation of sounds to achieve the desired effect or mood of audio signals using computer algorithms. This is significant for an extensive range of applications, including text-to-speech (TTS) systems and music generation [50]. Audio generation algorithms also known as vocoder in TTS and synthesizers in music respond to higher-level control signals to create fine-grained audio waveforms. Synthesizers have a long history of being hand designed instruments, accepting control signals and filter parameters to produce a tone. Audio synthesis uses both absolute and relative time series representation of features, and relevance feedback on both the feature weights and time series to refine the query [27].

Amplitude Modulation (AM) and Frequency Modulation (FM) synthesis are two traditional approaches that have enjoyed patronage over years in audio synthesis. These two methods usually involve the design and manual engineering of individual sound components; like oscillators and filters, and juxtaposing them in numerous ways to try to reproduce existing sounds or create completely new ones. While there has been some work done in the past on automatic parameter tuning for synthesizers, their potential success was limited to the few parameters they worked with. More recently, however, a ML approach has emerged, and its potential success was demonstrated in a recent paper published in [50]. It was revealed that over 300 million parameters can be extracted from audio signals and that DL can help discover what these core synthesis elements should be and how to connect them to get natural sounds.

Natural sounds have so many complexities and NNs are good at modeling these complexities, they further proposed that instead of getting a bunch of data, supervised learning DNNs could be trained on these huge parameters such that they automatically adjust themselves and be able to make natural sounds. This possibility was demonstrated with the duo; a WaveNet-style autoencoder (that learns temporal hidden codes to effectively capture longer-term structure without external conditioning) and NSynth, which is a large dataset for exploring neural audio synthesis of musical notes.

Journalism is also creative work as it requires critical thinking, judgment, storytelling, etc. However, as with all other domains, AI plays a key role in this particular field too. There are already robot journalists that generate articles and automate different processes in the media. Half of the content published by Bloomberg News uses automated flows. For instance, Cyborg is a system that helps a company compile a report every quarter. Forbes recently announced that they have tested a content management system called Bertie to provide reporters with early drafts and story templates. It is still very far from a robot that will write complete articles, since a lot of work is done at the back end, with editors and writers correcting several versions of the story. One of the benefits of journalist robots, however, as stated by Patch chief executive Warren St. John, is that there are no typos in the current articles written by AI.

DL is also revolutionizing the filmmaking industry in many ways. To reduce financial risks, movie revenue predictions have begun to be used in the industry [236]. Since there are already tools for generating music or paintings, creating movies is no exception. AI written movie called Sunspring was introduced by Oscar Sharp in 2016. Several hundred sci-fi screenplays from the 1980s and 90s were used as input to the LSTM RNN to generate its own version. It is not perfect yet, however, it is a big step towards making full use of AI in this industry and creating a movie on demand or making a movie per viewer's choice. Generating video from text is another challenge that can help achieve this goal based on DNN [123]. Other useful applications in cinematography are automated systems for subtitle generation and validation, that are achieved through extracting audio from video and then converting to text [14]. Large companies like Netflix are also using AI driven solutions to predict future demands at strategic server locations, allowing users to stream high-quality video even during peak hours.

With the growing availability of cheaper sensors and devices, greater innovation in hardware used in entertainment systems could be expected, rather than rapid innovation in software. For example, having virtual reality and haptics entering our rooms, or having personalized companion robots being our best friends. Leveraging AI, animators can create characters for VR games and movies. With improvements in automatic speech recognition, researchers expect that interactions with robots and other entertainment systems will become conversation-based. Interacting systems are expected to develop characteristics such as emotion, empathy, and adaptation to time, just like humans.

In the gaming industry, AI and DL rely on massive data sets to analyze and predict feature scenarios far beyond human capabilities. In 1997, IBM's Deep Blue defeated world chess champion Garry Kasparov, which was unrealistic at that time. In fact, the Deep Blue system relies on computational power rather than any actual learning [25]. Compared to Deep Blue, Google DeepMind created AlphaGO, which defeated world champion Lee Se-dol in Go who later announced his retirement due to AI [204]. Another AI-powered invention, AlphaStar, was trained to play StarCraft II. Here, NNs using reinforcement learning have been used to learn from previous experiences. However, in this case, the machine did not demonstrate the brilliant performance, as in the case of the AlphaGO. Consequently, more and more sophisticated AI and ML approaches are used every day to make machine opponents more engaging and harder to defeat.

Customer lifetime value prediction has become more popular with the advances of video games. Discount tactics for regulars and high rollers are widely used among casino managers. And so these techniques began to be used in the online market. CNN structures have proven to be most efficient at predicting the economic value of individual players, in other words, the amount of money they will spend on that video game during their lifetime. In [34], the authors proposed a 10 layered CNN model, including the input layer, the first convolutional layer, a max pooling layer, the second convolutional layer, the third convolutional layer, a flatten layer, three fully connected layers, and an output layer (Fig. 4.21). As shown in the paper, DNN approaches show higher accuracy and significantly better predictions for top spenders.

Every day more and more applications are used in social media and entertainment, ranging from assisting robots to opponent characters in games. Here

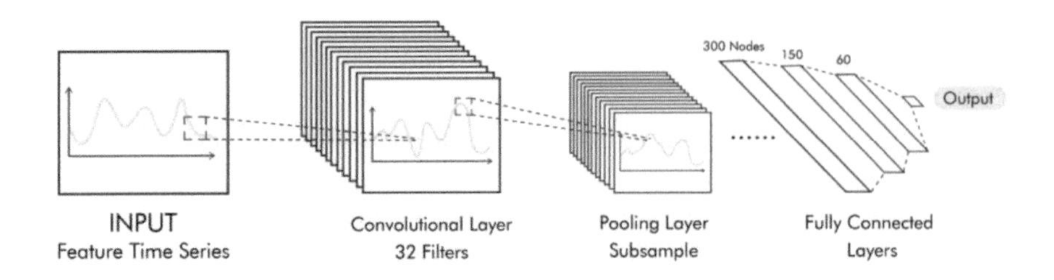

Figure 4.21: Structure of CNN [34].

are the main areas of these domains that have already been influenced by DNN solutions:

- Visual analysis and tagging,
- Social media listening and recommendations,
- Data transfer and compression,
- Sentiment analysis,
- Audio synthesis,
- Search optimization,
- Intelligent game opponents.

4.9 Transportation

AI has opened up incredible possibilities in various fields. This is especially true for the transportation industry. Autonomous transportation will soon get so widespread that it will most likely become the first experience with physically embodied AI systems for most people. This, in turn, will definitely influence the public's perception of AI as something as common as the Internet today. As cars will become better drivers than humans, people will eventually own fewer cars, live further away from work, and spend their time a lot more differently, leading to an entirely new urban structure. Furthermore, in 2030, in a typical North American city, it is expected that changes will not only be limited to cars and trucks, but will most likely also affect flying vehicles and personal robots.

As sensing algorithms achieve super-human performance for capabilities required for driving, automated perception, including vision, is currently approaching human-level performance for tasks such as recognition and tracking. Perception capabilities of machines will be followed by algorithmic improvements in higher level reasoning capabilities such as planning. The adoption of self-driving capabilities will not just be limited to personal transportation, but we will be expected to see self-driving and remotely controlled delivery vehicles, flying vehicles such as drones, and trucks. Ride sharing services are also likely to utilize self-driving vehicles. Advances in robotics will facilitate the creation of other types of autonomous vehicles, including robots and drones. With self-driving car technology, people will have more time to work or entertain themselves during their commutes. The increased comfort and decreased effort associated with self-driving cars and shared transportation may affect where people choose to live. The reduced need for parking may affect the way cities and public spaces are designed. Self-driving cars may also serve to increase the freedom and mobility of youth, elderly and people with disabilities. For example, there are already self-driving cars that may soon appear on the streets, such as the Olli electric shuttle from Local Motors (https://localmotors.com/meet-olli/) or TuSimple (https://www.tusimple.com/) whose driving system has been trained by DL to simulate tens of millions of miles of road driving.

NNs are used in autonomous vehicles to address various problems. For example, the first self-driving car, named ALVINN, used NNs to detect lane lines and segment the ground for driving [153]. Over the years, DL has begun to be used in the major sub-fields of self-driving vehicles. End-to-end autonomous driving systems typically use four blocks: sensors, perception, planning and control. The information obtained through various sensors on the vehicle is sent to a perception block, which is designed to detect obstacles and drivable space on the roads. The planning block uses meaningful information from perception for behavior and path planning. Finally, the control module checks to see if the vehicle is following the already defined paths or not. In [113], the authors have suggested an end-to-end light CNN solution for autonomous driving, called J-Net. The effectiveness of the proposed solution in comparison with existing methods is shown in Fig. 4.22.

As stated above, DL is the force that is bringing self-driving to life. Machines are trained to learn from millions of data sets from which a model is built, and only then they have been tested in a secure environment. There are other features that have been developed for driverless cars. For example, the Uber AI Labs added smart delivery options to their autonomous cars. The major concern for autonomous car companies is still remaining the unprecedented scenarios. The regular cycle of testing and implementation, typical to DL algorithms, ensures safe driving with more and more coverage of millions of scenarios. Data from cameras, sensors and geo-mapping help to create succinct and sophisticated models to navigate through traffic, identify paths, signage, pedestrian-only routes, and real-time elements like traffic volume and road blockages.

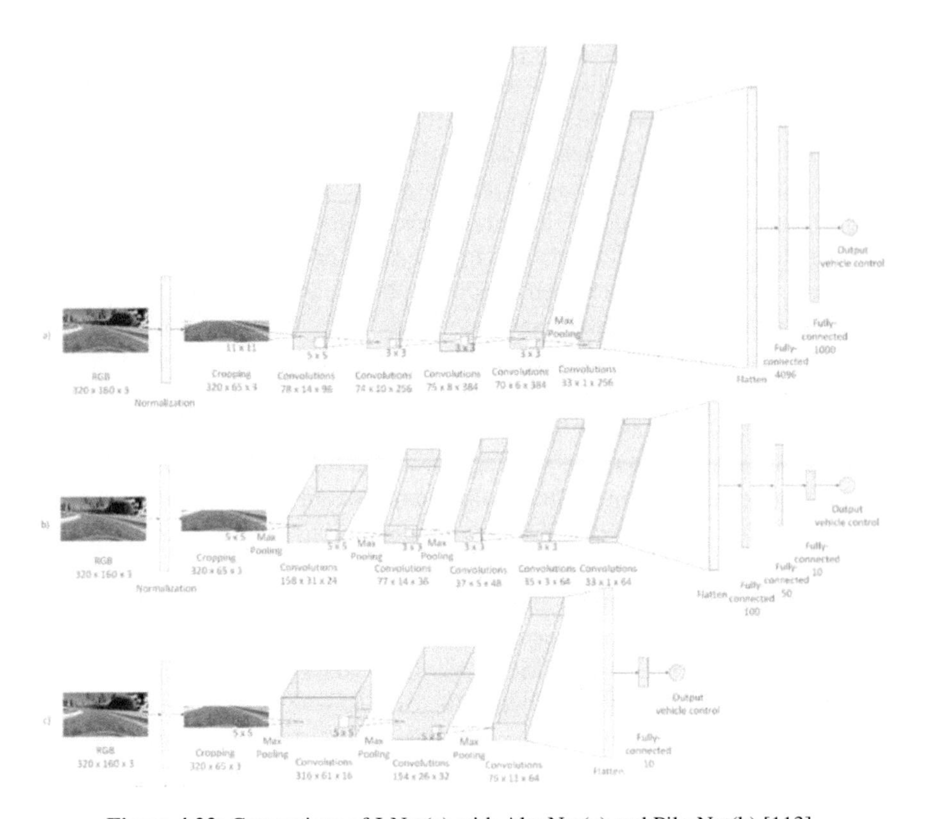

Figure 4.22: Comparison of J-Net(c) with AlexNet(a) and PilotNet(b) [113].

Cities around the world had begun investing in transportation infrastructure to develop sensing capabilities for vehicle and pedestrian traffic since 2005. Cities are using AI techniques to optimize services such as bus and mass rapid transit schedules and tracking traffic conditions to adjust speed limits or apply reasonable pricing on highways and bridges [221]. By using sensors and cameras in the road network, they can also optimize traffic light timing to improve traffic flow or predict traffic accidents and conditions [183]. These real-time strategies aim to conserve the limited resources of the transportation network, which is only made possible due to the availability of data and the widespread connectivity of individuals. One example of AI in transportation is Valerann smart road system (https://www.valerann.com/), which tracks road conditions and predicts highway situations using wireless sensors built into the roads.

In 2016, the United States Department of Transportation announced a call for proposals asking medium-size cities to imagine smart city infrastructure for transportation. They plan to award forty million dollars to a city that will successfully demonstrate how technology and data can be used to reform the movement of people as well as goods. One proposed solution is a network of connected vehicles that can reach an important level of safety and accuracy in driving through machine-to-machine communication. We can expect that advances in coordination and planning between multiple agents will have a significant impact on the vehicles of the future and play a role in improving the reliability and efficiency of the transportation system. Robots and drones are also likely to take part in transportation by transporting individuals and packaged goods. For transportation of goods, interest in drones has grown, and Amazon is currently testing its Prime Air delivery system. The well-known Vehicle Routing Problem (VRP) can be solved by applying ML/DL-powered route optimization software. This will help minimize the travel costs and the number of vehicles by creating optimal routes for various customers [93].

Another smart city infrastructure application is a traffic congestion prediction system that automatically monitors and minimizes the blockages in streets [15]. This type of real-time control system is possible by implementing various sensors on different adjacent signals that can collect, share traffic data and send to the NN for prediction through IoT enabled devices. On the other hand, for a long time, Google Maps is based on ML algorithms to suggest the best road to reach the endpoint. To predict the traffic situation in the near future, Google Maps analyzes historical traffic patterns on the streets over the time, achieving approximately 97% accuracy for all trips. However, it is important to note that these patterns have changed dramatically due to the COVID-19 pandemic. So, the models have been updated to automatically prioritize historical traffic patterns over the past four weeks.

Numerous traffic accidents have been reported due to distracted drivers, which is becoming a major problem in the transportation sector. Using data processing capabilities, DL can help here detect and analyze driver behavior. A warning can then be sent to the driver or, if repeated, an alert can then be sent directly to any officers within the local area to intercept them. In [9], the authors proposed a CNN architecture for detecting the distracted behavior (Fig. 4.23). They classify the input image into one of ten classes, including normal driving, texting while driving, using the right hand, talking to someone on the phone using the right hand, texting while driving using the left hand, talking to someone on the phone using the left hand, reaching for the dashboard to operate the radio, drinking or eating, reaching behind, fixing hair and makeup, and talking to a passenger.

Automation has been used in the aircraft industry for many years. The first autopilots date back to 1914. In modern airlines, autopilots perform all the work, they fly automatically until something unexpected happens, and only then the

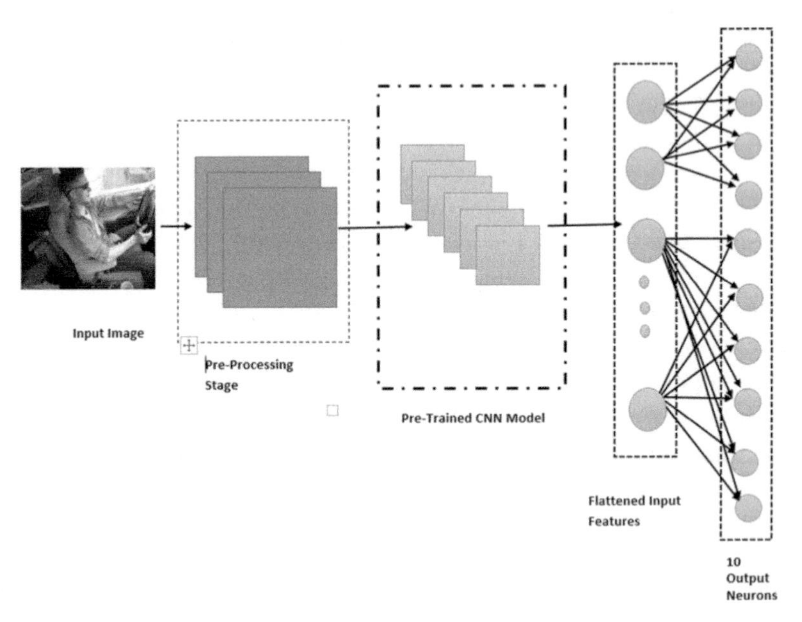

Figure 4.23: CNN architecture for distracted driver detection [9].

pilot takes over the control. AI applications are also included in the autopilot systems. However, most likely, in the near future, air transportation will not switch to a full autopilot system compared to the potential of self-driving cars. Nowadays, public acceptance of autonomous flight is only possible for single-passenger aircraft. Only the success of this model can lead to the adoption of fully autonomous systems for large passenger planes. To reduce dependencies on human pilots, the aviation industry has begun using various ML and DL algorithms to train the system to learn and act as its own. In recent years, several intelligent autopilot systems have been proposed that based on NNs, learn piloting skills from humans to perform a complete flight, including the autonomous actions such as landing and navigation [17]. It is important to collect all this knowledge in order to successfully handle the uncertainties. This could be the future of complete autopilot systems.

Flight delays are a critical issue in air transportation that directly impacts airline agency incomes and customer satisfaction. Due to the massive volume data and the extreme number of important parameters, only DL can help to more accurately predict flight delays. In [227], the authors proposed a deep graph-embedded LSTM model for predicting flight delays. Experiments were conducted at 325 major airports in United States for input signals such as departure

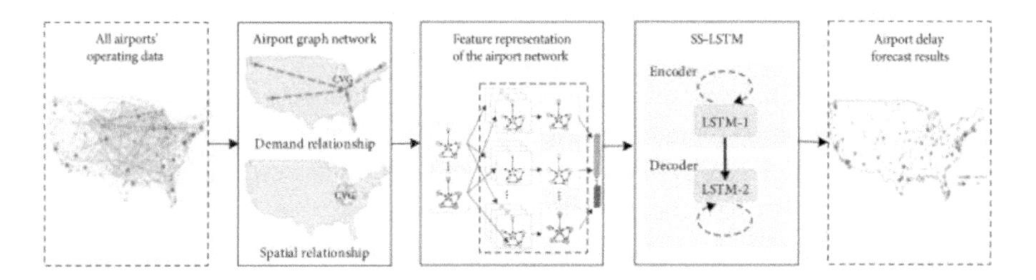

Figure 4.24: Deep graph-embedded LSTM model [227].

delay, arrival delay, departure flight volume, arrival flight volume, and airspace status.

The majority of companies in the transportation industry that rely on corrective and periodic maintenance strategies try to optimize costs by integrating automatic data-driven predictive maintenance models. Controlling the condition of vehicle to detect any anomalies or to determine the correct time for maintenance is important not only for drivers, but also for car manufacturers. Data mining algorithms have made it possible to predict engine stress, including also driver misbehavior, to help determine the need for vehicle maintenance in early stages [132]. To improve DNN performance for vehicle maintenance prediction, the geographic information system data can be used [33]. DL algorithms are also used to proactively predict highway maintenance problems, preventing expensive and unplanned failures on the roads. For example, the Cambridge-based AI company, Intellegens, announced a project that can successfully predict where drainage and gully blockages are most likely to occur in a busy road network, allowing them to be resolved before they become a problem.

AI solutions have long been used in the transportation sector. The most important areas affected by DL solutions are listed below:

- Self-driving vehicles,
- Traffic management and smart roads,
- Individual traffic planning,
- Air transportation,
- Connected transportation infrastructure,
- Law enforcement,
- Predictive fleet maintenance.

4.10 Other Applications

Since we covered various applications in previous industry sections, here we focus on the rest of the methods that are more general and not industry specific.

In engineering, DNN applications are more important, especially in highly reliable systems that have been used in various industry fields, including auto piloting systems, healthcare systems, automotive control, and others that require autonomy. AI, which is used in engineering, uses software and hardware components. As machines become more sophisticated, they will be able to support not only smart manufacturing tasks, but they will also be able to design and improve tasks over time without human intervention.

AI tools can provide automation for various tasks, freeing up engineers to perform more valuable work. By using DL solutions to discover patterns in data, machines can even help with computer engineering tasks. Code generation, also known as program synthesis, is a technique for producing executable forms of computer programs, such as machine code, in an automated fashion [20]. Several ML/DL techniques have been adopted to solve this task. The most recent implementation is the Neuro-Symbolic Program Synthesis (NSPS). A DL architecture for code generation was presented in [147]. It has been reported that the system once trained, can automatically generate computer programs that are consistent with any set of input-output examples provided during testing. It deploys two module neural architectures as its functional blocks: the cross-correlation I/O network that creates a continuous representation of any given set of input-output examples, and the Recursive-Reverse-Recursive Neural Network (R3NN), which synthesizes a program by incrementally building partial programs taking into account the continuous representation of the input-output examples.

Human capital has always been a big issue in the manufacturing industry. Due to the rise in the costs of living, as well as government regulations, costs of labor have increased significantly over the past few decades, resulting in increased production costs. Adoption of AI has reduced IT staff in the manufacturing industry by automating the procurement process. The yield has also been improved for integrated-circuit products when AI such as data integration and advanced analytics are used to improve R&D process. This is due to the effectiveness of AI-powered technologies, allowing to deliver more efficient designs by eliminating waste in the process, hence, increasing innovation delivery. On the other hand, predictive analysis integrated with DL solutions will significantly shorten the time required to solve design problems for semiconductor manufacturers by saving plenty of time for iterating and testing. For example, to make the production yield better, the engineers in semiconductor manufacturing processes can use NNs to predict machine outliers [222]. Fault detection and classification is an information system that can be used for this purpose, and it helps identify extreme parameters in a machine that is used for semiconductor processes. The growth of DL approaches promises the development of machines capable of performing more complex manufacturing and even design. Systems capable of learning without human intervention are also the ultimate goal of manufacturing processes, and this will have significant and far-reaching implications.

Microscopic flaws and defects can not be detected by the human eye without an error, but systems fitted with a powerful camera and inspection capabilities will be able to spot even the smallest defects. Deep CNNs are very common techniques used in computational lithography in semiconductor applications where nanometer level accuracy is important. Another application in semiconductor manufacturing is e-beam inspection, where DL solutions can be used to infer a smooth image captured by few repeated scans on the inspection machine. In [36], the authors have proposed a deep CNN model for detecting wafer surface defects. They achieved 98% accuracy when classifying 256x256 wafer images into four classes (center, local, random and scrape). The NN approach can be used for semiconductor wafer post-sawing inspection [187]. The overall conceptual schema of an NN-based inspection system is shown in Fig. 4.25. NN is used to process digitized die images extracted by electron microscope. These approaches help to reduce the inspection errors at various stages of the manufacturing process that may be introduced due to human fatigue.

The use of AI technologies untangles the procurement process to get a better grip on costs of production around the world. Keeping manufacturers stocked with parts is an extremely complex challenge. This is because thousands of various parts are sourced from thousands of suppliers around the world in order to ensure a smooth operation. Linking manufacturers digitally to their suppliers' systems enables transparency on supplier's availability and downtime. AI technologies can also help balance the supply chain and optimize inventories in real time. In the future, supplier research and analysis can be automated and optimized using ML/DL algorithms, e-auctions can be supplemented with virtual agents and in-person interactions can be programmed when needed. Similar to the retail industry, AI technologies can be applied to predict sales of maintenance services, optimizing pricing as well as refining sales-leads prioritization. In the future, manufacturers will be able to apply DL technology to optimize key performance indicators of program reviews in real time. Current DNNs are applied to make predictions based on past data. Therefore, tailoring of models in real time is key to deploying DL to real-time information that will better predict and prevent material and staffing bottlenecks and consumption. When natural language has been developed to a significant level, virtual agents can alert program leadership and team members and help solve problems. Factory managers can

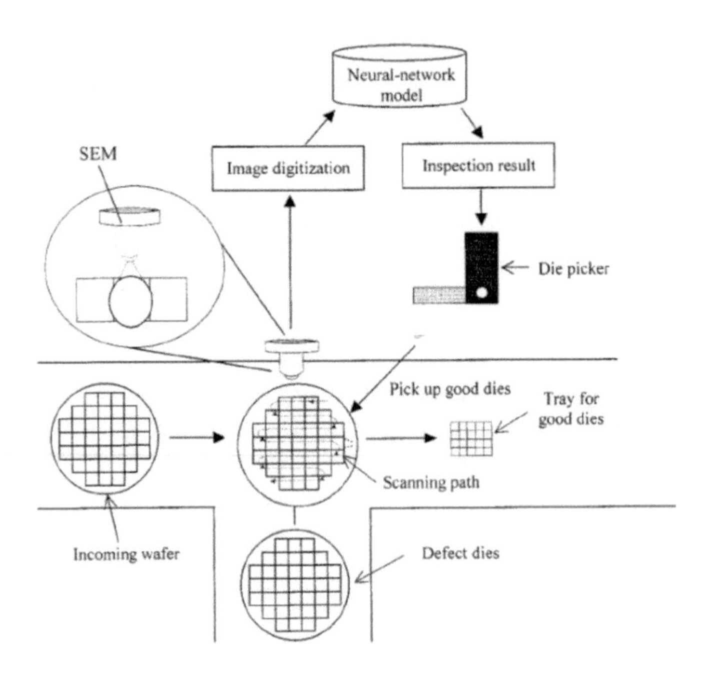

Figure 4.25: Conceptual schema for the NN based inspection system [187].

apply DL to real-time information flows to update and increase predictions with visibility on component availability and risk management. On the other hand, asset reliability can also be enhanced by using ML to improve the predictive accuracy of defaults.

The advancement of AI has allowed developers to create solutions that can carry out complex tasks in IoT systems. We have already discussed various DNN-based IoT applications in other sections. Here, we will cover remaining applications for smart homes and grids. In smart home systems, DL solutions are used to predict energy demand, posture detection, behavior monitoring, etc. In [124], the LSTM model was used to model spatio-temporal sequences that are required for smart home sensing systems. The LSTM model has been proven to be preferable over other existing ML approaches for this particular task. In order to have complete smart home systems, more supporting platforms have been developed in recent years. Cloud services can be used to collect available data from a smart environment, which can be used to generate NN models to predict energy demand and recommend improvements [154]. The RNN based approach has been shown to be a suitable choice for designing an intelligent forecasting model to predict energy demand in smart homes [139].

To achieve greater efficiency and reliability, power demand prediction mechanisms have been introduced in power grids. To achieve better accuracy for this task, a hybrid prediction model based on LSTM and CNN has been proposed [108]. The features used in this work include different types of contextual information such as temperature, humidity and season. These features define the number of LSTMs consisting of two hidden layers. The output of LSTMs is sent to CNN to predict power demand value. Similar to other fields, in power grid systems, DL solutions can be used to detect various anomalies [56].

AI has had an impact on just about every industry and sector. The following are fields that were not discussed in the previous sections:

- Computer engineering,
- Manufacturing and industrial engineering,
- Smart homes and grids.

Chapter 5

Discussions and Criticism

5.1 Challenges We Face Today

Advances in technology have always transformed the way human beings work and live throughout history. During the first industrial revolution, technology advancements included the shift from manual production processes to machine manufacturing, the introduction of chemicals, the iron production, the development of water power and steam power, and others. Coal was used as a primary power generating source. During the second industrial revolution, large scale manufacture of machine tools and railroads as well as large-scale steel and iron production emerged. Electricity and electrical communication were also invented. Petroleum, paper making machines, automobiles, maritime technology, use of chemicals, etc. were also developed. Income increased due to the increase in efficiency, and this, in turn, increased the standard of living for many people. Another century passes, and the third revolution has caused the rise of electronics, telecommunications and, of course, computers. Almost every aspect of people's lives has changed as a result of the three industrial revolutions.

The first industrial revolution benefited its leader of the time, Great Britain, by making it one of the strongest empires in the world, and, similarly, the second and third industrial revolutions created technologically and economically advanced superpowers such as the United States and Germany. And now the world is on the verge of fully ushering in the fourth industrial revolution, in which AI is the most important general-purpose technology. Like the steam engine that led to widespread commercial use of driving machineries in industries during the first industrial revolution; the internal combustion engine that gave rise to cars, trucks, and airplanes; electricity that caused the second industrial revolution through the discovery of direct and alternating current; and the Internet, which led to the emergence of the information age, AI is a transformational technology. This will cause a paradigm shift in the way problems are solved in every aspect of our lives, and, from it, innovative technologies will emerge. According to the head of the Toyota Research Institute, Gil Pratt, the current wave of AI technology can be compared with the Cambrian explosion 500 million years ago, which birthed a tremendous variety of new life forms. This new revolution is a truly global race. Countries today are racing to advance in AI, not only so that they won't be left behind, but more likely because historical trends have proven that whichever country conquers AI will conquer the world.

The AI applications that we have today have not even reached 1% of their potential. This is the reason why we are living in the most exciting period of the century, described by the World Economic Forum as the dawn of the fourth industrial revolution. "You cannot wait until a house burns down to buy fire insurance on it. We cannot wait until there are massive dislocations in our society to prepare for the Fourth Industrial Revolution." (Robert J. Shiller, 2013 Nobel laureate in economics, Professor of Economics, Yale University) History has shown that advances in technology are only part of the advancement of humanity, part of the inevitable. Just like any previous industrial revolution, technology advancement will significantly transform the way we live, work and interact with each other. Our dependencies today on mobile phones and the internet are great examples of how technologies transform us as a species. Despite bringing us conveniences, technology also greatly altered our needs and goals. The accumulation of adoption of different technologies and our subsequent needs and goals generated from the adoption, will make the world an entirely different place from where we are today.

What exactly incentivizes adoption of AI today? What are the exact reasons industries and, to a certain extent, governments embracing the adoption of AI in the first place? Increased profit margin is one of the highest incentives for industries to adopt AI technologies to automate their companies. A study by McKinsey Global Institute shows that firms that are currently combining strong digital capabilities, have robust AI adoptions, and proactive AI strategies see outsized financial performances. Firms adopting AI at scale or at the core of their businesses are already seeing the technology's potential, and those currently implementing proactive AI strategies are seeing even greater benefits in the long term.

The excitement and interest in DL is everywhere, capturing the imaginations of private companies, governments, and even patients. The impact of DNN applications on human lives and the current economy is inexplicable. According to the latest predictions, AI can boost business productivity by up to 40% depending on applications ranging from disease prediction to fraud detection. It is clear from the previous sections that DL applications are rapidly evolving across all industries. Along with the development of AI applications, the challenges we face today seem to be yet very serious, and we are still far from perfect systems. However, recognizing and working on these challenges can help propel the further growth of AI. Below, we take a look at the most common challenges that DL is facing today.

- Data availability and labeling—while humans seem to be able to generalize well with far less experience, AI systems require a thousand times more data to learn and recognize features. DNN applications are more popular in areas where data accessibility is not an issue. It will take time until DL solutions will become more efficient to be able to work with less data. On the other hand, for solutions that require labeled data, this becomes very challenging due to the unstructured origin of current IoT data.

- Flexibility—current DL is most successful at perception tasks and especially for those they have been trained for. The final goal is to make DL more flexible in solving various problems, by having some sort of knowledge of other similar problems it has faced before (meta-learning concept in ML). Hence, in order to have a general AI system, DNNs should learn multiple tasks. Progressive NNs are designed to connect several DL systems that will transfer certain bits of information, avoiding the learning process from the scratch [172]. However, there is still a long way to go to create a fully general and flexible AI system.

- Black box—how DL models come to certain decision usually remains unexplained, creating a lack of trust when working with them. The solution to resolve trust issues can be the creation of explainable AI.

- Computing power—modern DL models require a high number of cores and GPUs to run efficiently. Even with the availability of parallel processing and GPUs, it is still challenging due to the rapidly increasing number of complex algorithms and the unprecedented amount of data.

- Ethics and legal—DNN models predict outcome based on the data they have been trained on, ignoring the correctness of the data. If the model is trained on data that includes racism, the results will reflect that back without correcting them. This is why, from an ethical standpoint, an additional validation layer should be added to ensure the results are based on corrected data. On the other hand, if the AI system is using massive data and that data is sensitive, it might be in violation of federal laws. Therefore, when collecting data the organizations need to be careful even if the information is not harmless by itself, but is sensitive when collected together. With so much sensitive data, organizations need to be careful about the potential for data leaks.

- Accuracy and complexity of algorithms—the accuracy of current DNN models is still low (approximately 90–95%). DL models still require a lot of work to achieve high accuracy, including fine tuning, optimization of hyperparameters, large data sets, etc. Therefore, at present, continuous training of DL models might require significant manpower to support AI systems, which can become quite a challenge for organizations.

Despite the challenges outlined above, it is clear that AI has the potential to significantly boost the economy and help address some of the world's daunting challenges (e.g., climate change, healthcare, and education), by facilitating knowledge discovery in various domains of society. DNN applications today can learn to perform certain specific tasks, although they still lack real reasoning, commonsense knowledge, and understanding of concepts and language. Actions that have been carried out by humans in recent years are now partially or fully automated, and the list of actions continues to grow. The influence of AI will be magnified in the coming decade as manufacturing, retail, transportation, finance, healthcare, law, advertising, insurance, entertainment, education, and virtually every other industry is gearing to adopt it in transforming and improving their business processes and models.

5.2 Future of Deep Neural Networks

AI is the theory and development of machines that can imitate human intelligence in tasks such as visual perception, speech recognition, decision-making, and human language translation. Although AI has been around for several decades (now in the fourth or fifth wave), three factors have been identified as the major drivers of its recent growth and outburst:

- Data availability has increased as much as 1,000-fold. In fact, ninety percent of the digital data in the world today has been created in the past two years alone.

- Key algorithms have improved 10-fold to 100-fold performance-wise. The algorithms and approaches that now dominate the discipline, including deep supervised learning or reinforcement learning, share a vital basic property: their results improve with the size and amount of training data.
- The speed of hardware has improved by at least 100-fold. By a fortuitous coincidence, a type of computer chip called GPU turns out to be very effective when applied to the types of calculations needed by the robust algorithms. According to Shane Legg, a co-founder of Google DeepMind, training that takes one day on a single TPU device would have taken a quarter of a million years on an 80486 processor from 1990.

This growth of AI is currently bringing a lot of changes; and all these changes directly affect people, including the potential automation of jobs that have given people a sense of security and purpose over the past several centuries. There are various controversies when discussing the future of AI and its implications for humanity. Many have asked: "Will my job be taken away by robots?", and many people have had their minds on this question at least once, thinking whether we should welcome or be afraid of these changes.

AI experts from Oxford University and Yale University have conducted research to estimate when AI will exceed human performance in various jobs. One should not be too surprised about the automation of many low skilled and repetitive jobs, notably in manufacturing and logistics, as trends have shown that automation has already affected so many people. Experts were skeptical about whether it is possible for these low-skilled blue-collar jobs to be shipped back to America for one simple reason — automation. However, research conducted by experts shows that white collar jobs are not necessarily safe from automation either. Many predict that professions such as surgeons and lawyers may also be automated in the near future.

Employment has shifted due to a major recession and increasing globalization over the past decade. Enormous changes in non-AI digital technology have also been observed. The spectrum of tasks that digital systems can do is changing as rapidly as AI systems improve. Instead of affecting tasks historically performed by machines, AI is also creeping into the high end of the spectrum, including professional services. An understandable human fear of being marginalized by AI innovations will need to be overcome. However, AI is likely to replace tasks rather than jobs soon, and create new kinds of jobs. But the new jobs that will emerge are way speculated than the existing jobs that are likely to be lost. Since changes in employment usually happen gradually, often without a sharp transition, it is likely AI will slowly move to the workplace. The spectrum of effects is likely to emerge first, ranging from insignificant amounts of replacement or changes to an entire replacement. For example, AI applied to legal information extraction and topic modeling has almost automated parts of first-year lawyers' jobs. It is not difficult to predict that this will affect a diverse array of job-holders, from accountants to truck drivers to gardeners.

AI may also influence the size and location of the workforce. Many organizations and institutions are large because they rely on human labor to scale, either horizontally across geographical areas or vertically in management hierarchies. Since AI takes over many functions, scalability may no longer imply large organizations. Many have noticed the small number of employees of some high profile internet companies in recent years. The creation of efficiently outsourced labor markets brought upon by AI may enable enterprises towards that natural size where the CEO knows everyone. On the other hand, AI will also create jobs in some sectors, making certain tasks more important. AI can also create new categories of employment by making possible a new form of interaction. Well designed and produced information systems can be used to create new markets like Uber and Airbnb, which often have the effect of decreasing the level of barriers to entry and increasing participation. Much research has been conducted by the AI community to determine further ways of creating new markets and making existing ones operate more efficiently. While work has some form of intrinsic value to people, most people work to be able to purchase goods and services they value. Since AI systems can perform work that previously required human labor, they have the effect of lowering the cost of many goods and services, effectively making everyone richer. In addition, AI technologies have an immense potential to help address the needs of low-resource communities, given targeted incentives and funding priorities. Budding efforts are promising for now, and to counteract fears that AI may contribute to joblessness and other societal problems, AI may be able to provide solutions if communities are able to see AI being implemented to support communities with lower resources.

In addition to changes in philosophical concepts like work and life aside, humans can anticipate virtually free transportation due to autonomous vehicles and go wherever they like, whenever they want, as well as cheaper or even free healthcare and education. Education could be more diverse and flexible as opposed to rigid brick and mortar education centers we have. A university degree might not be necessary to land a well-paying job, as people are free to pursue whichever field they want at their own pace. Innovations and creativity could be pursued by many, and advancement in various fields can be accelerated with increased enthusiasm and passion in the human workforce.

The fact that AI might be able to assist in spreading health-related information shows that social networks can be a great tool to be harnessed to create earlier, less-costly interventions involving large populations. Individual interventions are difficult and expensive, especially in areas like Los Angeles, where the number of homeless youths is beyond 5000 people, where the youths' mistrust of authorities dictates that key messages are best spread through peer leaders. Peer leaders can be selected with AI programs so that they will be able to leverage homeless youth social networks to strategically spread health-related information, such as how to avoid the spread of HIV. However, the dynamic, uncertain nature of these networks does pose challenges for AI research. Care must also be taken to avoid AI systems to reproduce discriminatory behavior, such as ML that identifies people through illegal racial indicators, or through highly-correlated surrogate factors, such as postal codes. However, if great care and regulations are applied, greater reliance on AI may well result in a reduction in discrimination overall, since AI programs are inherently more easily audited than humans.

The entertainment industry is the one that has benefited greatly from AI in recent years and is still growing towards satisfying the preferences of every user. In the near future, it will be possible to order a custom movie with the virtual actors of your preferences. The fast growth of medical applications, in which AI will drive a personalized medicine revolution, is already becoming a reality. AI algorithms are already helping doctors better analyze data and customize the health care system for each individual patient. Soon, with the help of brain tumor diagnoses, it will be possible to decide which cancer treatment will work best for a patient.

In order to achieve these predictions, the challenges connected to DL need to be resolved soon. As shown in the previous section, while DL systems generally become better and more powerful as they scale up, there are still fundamental deficiencies of current systems that cannot be overcome by scaling alone. One approach to close the gap between AI and human intelligence is hybrid AI that combines NN models with classical symbolic systems, which is still a major challenge for modern DL systems. Predictions for the DL future claim that DL tools will become a standard part of the software toolkit over the next five years. In addition, neural architecture search may also play a key role in building data sets for DL models. These enhancements will speed up the learning process of the models by several times.

Another major fear humanity is facing due to advances in AI is ethical and security issues. People are afraid that these systems, based on logic and programmed algorithms without any emotion, could harm them in order to achieve their own goals. For example, during military operations, logic cannot be the only determinant. And what about the consequences of such an error? When the AI system makes a mistake, the situation is a little more sophisticated. A super intelligent AI will be extremely good at achieving its goals, and this can become a problem for humanity. While everyone expects fantastic things from AI, there is no reason to be so fatalistic. How long will it take for machines to get human level intelligence? There are many discussions and predictions on this topic that this will happen in the next ten years. However, this is a common misconception, and it is hard to do any forecasting. While no one can exactly predict how DNN applications will evolve in the near future, existing models show that they are becoming an important part of our lives lately. Yes nowadays, these "narrow" AI models are everywhere, embedded in our smart phones and search engines. However, it is important to note that human-level intelligence must be an inspiration for the future of AI, but not something we should be trying to replicate. The future of DL is heading towards allowing DNNs to be more flexible and quickly adapt to new environments. Recent advances, including Transformers, Recurrent Independent Mechanisms, Meta-learning and more, are going in this direction, making AI more exciting.

Since three industrial revolutions have transformed the world during the 18th and 19th centuries, we are now living in a decade of AI revolution where DNN applications are potentially limitless. And we should accept that this revolution is having a positive impact on the world. DL with AI has already opened incredible possibilities in various fields. With an extremely high number of use cases and a growing amount of data to support these technologies, DL will no doubt play a key role in delivering the highest quality care to consumers in the very near future. As renowned computer scientist Alan Kay once said, "The only way to predict the future is to invent it." In order to truly know how our lives will be in the coming decades, now is the time to be onboard on the journey towards our next technological revolution, which is no doubt exciting and extremely promising!

References

[1] Emerson Rodolfo Abraham, João Gilberto Mendes dos Reis, Oduvaldo Vendrametto, Pedro Luiz de Oliveira Costa Neto, Rodrigo Carlo Toloi, Aguinaldo Eduardo de Souza, and Marcos de Oliveira Morais. Time series prediction with artificial neural networks: An analysis using brazilian soybean production. *Agriculture*, 10(10): 475, 2020.

[2] Roger Achkar, and Michel Owayjan. Implementation of a vision system for a landmine detecting robot using artificial neural network. *arXiv preprint arXiv:1210.7956*, 2012.

[3] Tarun Agrawal, and Prakash Choudhary. Focuscovid: Automated covid-19 detection using deep learning with chest x-ray images. *Evolving Systems*, 1–15, 2021.

[4] Igor Aizenberg, Naum N Aizenberg, and Joos PL Vandewalle. *Multivalued and Universal Binary Neurons: Theory, Learning and Applications*. Springer Science & Business Media, 2013.

[5] Simeon Okechukwu Ajakwe, Cosmas Ifeanyi Nwakanma, Jae-Min Lee, and Dong-Seong Kim. Machine learning algorithm for intelligent prediction for military logistics and planning. In *2020 International Conference on Information and Communication Technology Convergence (ICTC)*, pp. 417–419. IEEE, 2020.

[6] Khalid Al-Jabery, Tayo Obafemi-Ajayi, Gayla Olbricht, and Donald Wunsch. *Computational Learning Approaches to Data Analytics in Biomedical Applications*. Academic Press, 2019.

[7] David Alaminos, Rafael Becerra-Vicario, Manuel Á Fernández-Gámez, and Ana J Cisneros Ruiz. Currency crises prediction using deep neural decision trees. *Applied Sciences*, 9(23): 5227, 2019.

[8] Alammar. The illustrated transformer. Blog, 2018.

[9] Khalid A AlShalfan, and Mohammed Zakariah. Detecting driver distraction using deep-learning approach. CMC-Computers Materials & Continua, 68(1): 689–704, 2021.

[10] Luigi Ambrosio, and Gianni Dal Maso. A general chain rule for distributional derivatives. *Proceedings of the American Mathematical Society*, 108(3): 691–702, 1990.

[11] Boaz Arad, Jos Balendonck, Ruud Barth, Ohad Ben-Shahar, Yael Edan, Thomas Hellström, Jochen Hemming, Polina Kurtser, Ola Ringdahl, Toon Tielen, and Bart van Tuijl. Development of a sweet pepper harvesting robot. *Journal of Field Robotics*, 37(6): 1027–1039, 2020.

[12] Ashay Argal, Siddharth Gupta, Ajay Modi, Pratik Pandey, Simon Shim, and Chang Choo. Intelligent travel chatbot for predictive recommendation in echo platform. In *2018 IEEE 8th annual computing and communication workshop and conference (CCWC)*, pp. 176–183. IEEE, 2018.

[13] Sercan Ö. Arık, Mike Chrzanowski, Adam Coates, Gregory Diamos, Andrew Gibiansky, Yongguo Kang, Xian Li, John Miller, Andrew Ng, Jonathan Raiman, Shubho Sengupta, and Mohammad Shoeybi. Deep voice: Real-time neural text-to-speech. In *International Conference on Machine Learning*, pp. 195–204. PMLR, 2017.

[14] VB Aswin, Mohammed Javed, Parag Parihar, K Aswanth, CR Druval, Anupam Dagar, and CV Aravinda. Nlp-driven ensemble-based automatic subtitle generation and semantic video summarization technique. In *Advances in Artificial Intelligence and Data Engineering*, pp. 3–13. Springer, 2021.

[15] Ayesha Ata, Muhammad Adnan Khan, Sagheer Abbas, Gulzar Ahmad, and Areej Fatima. Modelling smart road traffic congestion control system using machine learning techniques. *Neural Network World*, 29(2): 99–110, 2019.

[16] Pedro Ballester, and Ricardo Araujo. On the performance of googlenet and alexnet applied to sketches. In *Proceedings of the AAAI Conference on Artificial Intelligence*, volume 30, 2016.

[17] Haitham Baomar, and Peter J Bentley. Autonomous navigation and landing of large jets using artificial neural networks and learning by imitation. In *2017 IEEE Symposium Series on Computational Intelligence (SSCI)*, pp. 1–10. IEEE, 2017.

[18] Ghazal Bargshady, Jeffrey Soar, Xujuan Zhou, Ravinesh C Deo, Frank Whittaker, and Hua Wang. A joint deep neural network model for pain recognition from face. In *2019 IEEE 4th International Conference on Computer and Communication Systems (ICCCS)*, pp. 52–56. IEEE, 2019.

[19] William Bechtel, and Adele Abrahamsen. *Connectionism and the Mind: An Introduction to Parallel processing in Networks*. Basil Blackwell, 1991.

[20] Kory Becker, and Justin Gottschlich. Ai programmer: Autonomously creating software programs using genetic algorithms. In *Proceedings of the Genetic and Evolutionary Computation Conference Companion*, pp. 1513–1521, 2021.

[21] Michael Beetz, Martin Buss, and Dirk Wollherr. Cognitive technical systems—what is the role of artificial intelligence? In *Annual Conference on Artificial Intelligence*, pp. 19–42. Springer, 2007.

[22] Bernhard E Boser, Isabelle M Guyon, and Vladimir N Vapnik. A training algorithm for optimal margin classifiers. In *Proceedings of the Fifth Annual Workshop on Computational Learning Theory*, pp. 144–152, 1992.

[23] Leo Breiman. Random forests. *Machine Learning*, 45(1): 5–32, 2001.

[24] Andrzej Cader. The potential for the use of deep neural networks in e-learning student evaluation with new data augmentation method. In *International Conference on Artificial Intelligence in Education*, pp. 37–42. Springer, 2020.

[25] Murray Campbell, A Joseph Hoane Jr, and Feng-hsiung Hsu. Deep blue. *Artificial Intelligence*, 134(1-2): 57–83, 2002.

[26] Flavio Capraro, Daniel Patino, Santiago Tosetti, and Carlos Schugurensky. Neural network-based irrigation control for precision agriculture. In *2008 IEEE International Conference on Networking, Sensing and Control*, pp. 357–362. IEEE, 2008.

[27] Mark Cartwright, and Bryan Pardo. Synthassist: An audio synthesizer programmed with vocal imitation. In *Proceedings of the 22nd ACM International Conference on Multimedia*, pp. 741–742, 2014.

[28] Ilias Chalkidis, Ion Androutsopoulos, and Nikolaos Aletras. Neural legal judgment prediction in english. *arXiv preprint arXiv:1906.02059*, 2019.

[29] Ellison Chan, Adam Krzyzak, and Ching Y Suen. Predicting us elections with social media and neural networks. In *International Conference on Pattern Recognition and Artificial Intelligence*, pp. 325–335. Springer, 2020.

[30] Chih-Chung Chang, and Chih-Jen Lin. Libsvm: A library for support vector machines. *ACM Transactions on Intelligent Systems and Technology (TIST)*, 2(3): 1–27, 2011.

[31] Nitin Kumar Chauhan, and Krishna Singh. A review on conventional machine learning vs deep learning. In *2018 International Conference on Computing, Power and Communication Technologies (GUCON)*, pp. 347–352. IEEE, 2018.

[32] Thira Chavarnakul, and David Enke. Intelligent technical analysis based equivolume charting for stock trading using neural networks. *Expert Systems with Applications*, 34(2): 1004–1017, 2008.

[33] Chong Chen, Ying Liu, Xianfang Sun, Carla Di Cairano-Gilfedder, and Scott Titmus. Automobile maintenance prediction using deep learning with gis data. *Procedia CIRP*, 81: 447–452, 2019.

[34] Pei Pei Chen, Anna Guitart, Ana Fernández del Río, and Africa Periánez. Customer lifetime value in video games using deep learning and parametric models. In *2018 IEEE International Conference on Big Data (Big Data)*, pp. 2134–2140. IEEE, 2018.

[35] I-Tsun Chiang, Jong-Chang Tsai, and Shang-Ti Chen. Using xbox 360 kinect games on enhancing visual performance skills on institutionalized

older adults with wheelchairs. In *2012 IEEE Fourth International Conference on Digital Game and Intelligent Toy Enhanced Learning*, pp. 263–267. IEEE, 2012.

[36] Jong-Chih Chien, Ming-Tao Wu, and Jiann-Der Lee. Inspection and classification of semiconductor wafer surface defects using CNN deep learning networks. *Applied Sciences*, 10(15): 5340, 2020.

[37] Yoonsuck Choe, Cesare Alippi, Robert Kozma, and Francesco Carlo Morabito. *Artificial Intelligence in the Age of Neural Networks and Brain Computing*. Academic Press, 2019.

[38] Ryan Dahl, Mohammad Norouzi, and Jonathon Shlens. Pixel recursive super resolution. In *Proceedings of the IEEE International Conference on Computer Vision*, pp. 5439–5448, 2017.

[39] Wenyuan Dai, Gui-Rong Xue, Qiang Yang, and Yong Yu. Co-clustering based classification for out-of-domain documents. In *Proceedings of the 13th ACM SIGKDD International Conference on Knowledge Discovery and Data Mining*, pp. 210–219, 2007.

[40] Bharath Dandala, Venkata Joopudi, and Murthy Devarakonda. Adverse drug events detection in clinical notes by jointly modeling entities and relations using neural networks. *Drug Safety*, 42(1): 135–146, 2019.

[41] Yann Dauphin, Razvan Pascanu, Caglar Gulcehre, Kyunghyun Cho, Surya Ganguli, and Yoshua Bengio. Identifying and attacking the saddle point problem in high-dimensional non-convex optimization. *arXiv preprint arXiv:1406.2572*, 2014.

[42] Alexander Jakob Dautel, Wolfgang Karl Härdle, Stefan Lessmann, and Hsin-Vonn Seow. Forex exchange rate forecasting using deep recurrent neural networks. *Digital Finance*, 2(1): 69–96, 2020.

[43] Jeffrey De Fauw, Joseph R Ledsam, Bernardino Romera-Paredes, Stanislav Nikolov, Nenad Tomasev, Sam Blackwell, Harry Askham, Xavier Glorot, Brendan O'Donoghue, Daniel Visentin, George van den Driessche, Balaji Lakshminarayanan, Clemens Meyer, Faith Mackinder, Simon Bouton, Kareem Ayoub, Reena Chopra, Dominic King, Alan Karthikesalingam, Cían O Hughes, Rosalind Raine, Julian Hughes, Dawn A Sim, Catherine Egan, Adnan Tufail, Hugh Montgomery, Demis Hassabis, Geraint Rees,

Trevor Back, Peng T Khaw, Mustafa Suleyman, Julien Cornebise, Pearse A Keane, and Olaf Ronneberger. Clinically applicable deep learning for diagnosis and referral in retinal disease. *Nature Medicine*, 24(9): 1342–1350, 2018.

[44] Jacob Devlin, Ming-Wei Chang, Kenton Lee, and Kristina Toutanova. Bert: Pre-training of deep bidirectional transformers for language understanding. *arXiv preprint arXiv:1810.04805*, 2018.

[45] Luke Dormehl. *Thinking Machines: The Quest for Artificial Intelligence—and Where It's Taking Us Next*. Penguin, 2017.

[46] John Duchi, Elad Hazan, and Yoram Singer. Adaptive subgradient methods for online learning and stochastic optimization. Journal of Machine Learning Research, 12(7), 2011.

[47] Atul Kumar Dwivedi, and Babita Sahu. Enhanced image object recognition system using correlation filter based on optimization. 2017.

[48] Eklavya. Kohonen self-organizing maps. a special type of artificial neural network. Webpage, 2019.

[49] Abdelrafe Elzamly, Burairah Hussin, Samy S Abu-Naser, Tadahiro Shibutani, and Mohamed Doheir. Predicting critical cloud computing security issues using artificial neural network (anns) algorithms in banking organizations. 2017.

[50] Jesse Engel, Cinjon Resnick, Adam Roberts, Sander Dieleman, Mohammad Norouzi, Douglas Eck, and Karen Simonyan. Neural audio synthesis of musical notes with wavenet autoencoders. In *International Conference on Machine Learning*, pp. 1068–1077. PMLR, 2017.

[51] Steve Engels, Vivek Lakshmanan, and Michelle Craig. Plagiarism detection using feature-based neural networks. In *Proceedings of the 38th SIGCSE Technical Symposium on Computer Science Education*, pp. 34–38, 2007.

[52] Andre Esteva, Brett Kuprel, Roberto A Novoa, Justin Ko, Susan M Swetter, Helen M Blau, and Sebastian Thrun. Dermatologist-level classification of skin cancer with deep neural networks. *Nature*, 542(7639): 115–118, 2017.

[53] Kawin Ethayarajh. How contextual are contextualized word representations? Comparing the geometry of bert, elmo, and gpt-2 embeddings. *arXiv preprint arXiv:1909.00512*, 2019.

[54] Farragher. Here are the mind-blowing things a deconvolutional neural network can do. Webpage, 2019.

[55] Chengwei Fei, Rong Liu, Zihao Li, Tianmin Wang, and Faisal N Baig. Machine and deep learning algorithms for wearable health monitoring. *Computational Intelligence in Healthcare*, 105.

[56] Zheng Fengming, Li Shufang, Guo Zhimin, Wu Bo, Tian Shiming, and Pan Mingming. Anomaly detection in smart grid based on encoder-decoder framework with recurrent neural network. *The Journal of China Universities of Posts and Telecommunications*, 24(6): 67–73, 2017.

[57] Rostislav Fojtik. The ozobot and education of programming. *New Trends and Issues Proceedings on Humanities and Social Sciences*, 4(5), 2017.

[58] Lubna A Gabralla, Rania Jammazi, and Ajith Abraham. Oil price prediction using ensemble machine learning. In *2013 International Conference on Computing, Electrical and Electronic Engineering (ICCEEE)*, pp. 674–679. IEEE, 2013.

[59] Jing Gao, Wei Fan, Jing Jiang, and Jiawei Han. Knowledge transfer via multiple model local structure mapping. In *Proceedings of the 14th ACM SIGKDD International Conference on Knowledge Discovery and Data Mining*, pp. 283–291, 2008.

[60] Sheila Garfield, and Stefan Wermter. Call classification using recurrent neural networks, support vector machines and finite state automata. *Knowledge and Information Systems*, 9(2): 131–156, 2006.

[61] Sushmito Ghosh, and Douglas L Reilly. Credit card fraud detection with a neural-network. In *System Sciences, 1994. Proceedings of the Twenty-Seventh Hawaii International Conference on*, volume 3, pp. 621–630. IEEE, 1994.

[62] Xavier Glorot, and Yoshua Bengio. Understanding the difficulty of training deep feedforward neural networks. In *Proceedings of the Thirteenth*

International Conference on Artificial Intelligence and Statistics, pp. 249–256. JMLR Workshop and Conference Proceedings, 2010.

[63] Ian Goodfellow. Nips 2016 tutorial: Generative adversarial networks. *arXiv preprint arXiv:1701.00160*, 2016.

[64] Ian Goodfellow, Yoshua Bengio, and Aaron Courville. Deep learning (adaptive computation and machine learning series). *DOI*, 10: 1762–1766, 2016.

[65] Ian J Goodfellow, Jean Pouget-Abadie, Mehdi Mirza, Bing Xu, David Warde-Farley, Sherjil Ozair, Aaron Courville, and Yoshua Bengio. Generative adversarial networks. *arXiv preprint arXiv:1406.2661*, 2014.

[66] Alex Graves, Abdel-rahman Mohamed, and Geoffrey Hinton. Speech recognition with deep recurrent neural networks. In *2013 IEEE International Conference on Acoustics, Speech and Signal Processing*, pp. 6645–6649. IEEE, 2013.

[67] Werner H Greub. *Linear Algebra*, volume 23. Springer Science & Business Media, 2012.

[68] Bo Guo, Rui Zhang, Guang Xu, Chuangming Shi, and Li Yang. Predicting students performance in educational data mining. In *2015 International Symposium on Educational Technology (ISET)*, pp. 125–128. IEEE, 2015.

[69] Hangzhi Guo, Alexander Woodruff, and Amulya Yadav. Improving lives of indebted farmers using deep learning: Predicting agricultural produce prices using convolutional neural networks. In *Proceedings of the AAAI Conference on Artificial Intelligence*, volume 34, pp. 13294–13299, 2020.

[70] Aaryan Gupta, Vinya Dengre, Hamza Abubakar Kheruwala, and Manan Shah. Comprehensive review of text-mining applications in finance. *Financial Innovation*, 6(1): 1–25, 2020.

[71] Akshay Kumar Gupta. Survey of visual question answering: Datasets and techniques. *arXiv preprint arXiv:1705.03865*, 2017.

[72] Michael Haenlein and Andreas Kaplan. A brief history of artificial intelligence: On the past, present, and future of artificial intelligence. *California Management Review*, 61(4): 5–14, 2019.

[73] Seung Seog Han, Ik Jun Moon, Woohyung Lim, In Suck Suh, Sam Yong Lee, Jung-Im Na, Seong Hwan Kim, and Sung Eun Chang. Keratinocytic skin cancer detection on the face using region-based convolutional neural network. *JAMA Dermatology*, 156(1): 29–37, 2020.

[74] Meliha Handzic, Felix Tjandrawibawa, and Julia Yeo. How neural networks can help loan officers to make better informed application decisions. *Informing Science*, 6: 97–109, 2003.

[75] Nga Tran Anh Hang. Applying deep neural network to retrieve relevant civil law articles. In *Proceedings of the Student Research Workshop Associated with RANLP*, pp. 46–48, 2017.

[76] Mohamad H Hassoun. *Fundamentals of Artificial Neural Networks*. MIT Press, 1995.

[77] Kaiming He, Xiangyu Zhang, Shaoqing Ren, and Jian Sun. Delving deep into rectifiers: Surpassing human-level performance on image net classification. In *Proceedings of the IEEE International Conference on Computer Vision*, pp. 1026–1034, 2015.

[78] Kaiming He, Xiangyu Zhang, Shaoqing Ren, and Jian Sun. Deep residual learning for image recognition. In *Proceedings of the IEEE Conference on Computer Vision and Pattern Recognition*, pp. 770–778, 2016.

[79] Jeff Heaton. *Introduction to Neural Networks with Java*. Heaton Research, Inc., 2008.

[80] Antonio Hernández-Blanco, Boris Herrera-Flores, David Tomás, and Borja Navarro-Colorado. A systematic review of deep learning approaches to educational data mining. *Complexity*, 2019.

[81] Rob High. The era of cognitive systems: An inside look at ibm watson and how it works. *IBM Corporation, Redbooks*, 1: 16, 2012.

[82] Geoffrey E Hinton. Deep belief networks. *Scholarpedia*, 4(5): 5947, 2009.

[83] Taekeun Hong, Jin-A Choi, Kiho Lim, and Pankoo Kim. Enhancing personalized ads using interest category classification of sns users based on deep neural networks. *Sensors*, 21(1): 199, 2021.

[84] David W Hosmer Jr, Stanley Lemeshow, and Rodney X Sturdivant. *Applied Logistic Regression*, volume 398. John Wiley & Sons, 2013.

[85] Tianyang Hu, Wenjia Wang, Cong Lin, and Guang Cheng. Regularization matters: A nonparametric perspective on over parametrized neural network. In *International Conference on Artificial Intelligence and Statistics*, pp. 829–837. PMLR, 2021.

[86] Wen-Chen Hu, Yanjun Zuo, Naima Kaabouch, and Lei Chen. An optimization neural network for smartphone data protection. In *2010 IEEE International Conference on Electro/Information Technology*, pp. 1–6. IEEE, 2010.

[87] Han-Chen Huang. Construction of a health food demand prediction model using a back propagation neural network. *Advance Journal of Food Science and Technology*, 5(7): 896–899, 2013.

[88] David H Hubel, and Torsten N Wiesel. Receptive fields, binocular interaction and functional architecture in the cat's visual cortex. *The Journal of Physiology*, 160: 106–154, 1962.

[89] Helenca Duxans i Barrobes. Voice conversion applied to text-to-speech systems. *Universitat Politecnica de Catalunya*, 2006.

[90] Sergey Ioffe and Christian Szegedy. Batch normalization: Accelerating deep network training by reducing internal covariate shift. In *International Conference on Machine Learning*, pp. 448–456. PMLR, 2015.

[91] Hafzullah İş, and Taner Tuncer. Interaction-based behavioral analysis of twitter social network accounts. *Applied Sciences*, 9(20): 4448, 2019.

[92] Félix J López Iturriaga, and Iván Pastor Sanz. Bankruptcy visualization and prediction using neural networks: A study of us commercial banks. *Expert Systems with Applications*, 42(6): 2857–2869, 2015.

[93] Waldy Joe, and Hoong Chuin Lau. Deep reinforcement learning approach to solve dynamic vehicle routing problem with stochastic customers. In *Proceedings of the International Conference on Automated Planning and Scheduling*, volume 30, pp. 394–402, 2020.

[94] Sérgio Jorge, Carlos Pereira, and Paulo Novais. Intelligent call routing for telecommunications call-centers. In *International Conference on*

Intelligent Data Engineering and Automated Learning, pp. 316–328. Springer, 2020.

[95] Rafal Jozefowicz, Wojciech Zaremba, and Ilya Sutskever. An empirical exploration of recurrent network architectures. In *International Conference on Machine Learning*, pp. 2342–2350. PMLR, 2015.

[96] Mohamed E Karar, MF Al-Rasheed, AF Al-Rasheed, and Omar Reyad. Iot and neural network-based water pumping control system for smart irrigation. *arXiv preprint arXiv:2005.04158*, 2020.

[97] Andrej Karpathy, and Li Fei-Fei. Deep visual-semantic alignments for generating image descriptions. In *Proceedings of the IEEE Conference on Computer Vision and Pattern Recognition*, pp. 3128–3137, 2015.

[98] S Karthik, R Srinivasa Perumal, and PVSSR Chandra Mouli. Breast cancer classification using deep neural networks. In *Knowledge Computing and its Applications*, pp. 227–241. Springer, 2018.

[99] Chaitanya Kaul, Suresh Manandhar, and Nick Pears. Focusnet: An attention-based fully convolutional network for medical image segmentation. In *2019 IEEE 16th International Symposium on Biomedical Omaging (ISBI 2019)*, pp. 455–458. IEEE, 2019.

[100] Matcheri S Keshavan, and Mukund Sudarshan. Deep dreaming, aberrant salience and psychosis: Connecting the dots by artificial neural networks. *Schizophrenia Research*, 188: 178–181, 2017.

[101] Cem Keskin, Furkan Kıraç, Yunus Emre Kara, and Lale Akarun. Real time hand pose estimation using depth sensors. In *Consumer Depth Cameras for Computer Vision*, pp. 119–137. Springer, 2013.

[102] Saeed Khaki, and Lizhi Wang. Crop yield prediction using deep neural networks. *Frontiers in Plant Science*, 10: 621, 2019.

[103] Hafsa Khalid, Muzammil Hussain, Mohammed A. Al Ghamdi, Tayyaba Khalid, Khadija Khalid, Muhammad Adnan Khan, Kalsoom Fatima, Khalid Masood, Sultan H. Almotiri, Muhammad Shoaib, and Aqsa Ahmed. A comparative systematic literature review on knee bone reports from mri, X-rays and CT scans using deep learning and machine learning methodologies. *Diagnostics*, 10(8): 518, 2020.

[104] Jibran Rasheed Khan, Muhammad Saeed, Farhan Ahmed Siddiqui, Nadeem Mahmood, and Qamar Ul Arifeen. Predictive policing: A machine learning approach to predict and control crimes in metropolitan cities. *University of Sindh Journal of Information and Communication Technology*, 3(1): 17–26, 2019.

[105] Muhammad Jaleed Khan, Adeel Yousaf, Nizwa Javed, Shifa Nadeem, and Khurram Khurshid. Automatic target detection in satellite images using deep learning. *Journal of Space Technology*, 7(1): 44–49, 2017.

[106] A Khashman. Automatic detection of military targets utilising neural networks and scale space analysis. Technical report, Near East Univ. Mersin (Turkey) Dept. of Computer Science, 2001.

[107] Sarit Khirirat, Hamid Reza Feyzmahdavian, and Mikael Johansson. Minibatch gradient descent: Faster convergence under data sparsity. In *2017 IEEE 56th Annual Conference on Decision and Control (CDC)*, pp. 2880–2887. IEEE, 2017.

[108] Myoungsoo Kim, Wonik Choi, Youngjun Jeon, and Ling Liu. A hybrid neural network model for power demand forecasting. *Energies*, 12(5): 931, 2019.

[109] Takashi Kimoto, Kazuo Asakawa, Morio Yoda, and Masakazu Takeoka. Stock market prediction system with modular neural networks. In *1990 IJCNN International Joint Conference on Neural Networks*, pp. 1–6. IEEE, 1990.

[110] Diederik P Kingma, and Jimmy Ba. Adam: A method for stochastic optimization. *arXiv preprint arXiv:1412.6980*, 2014.

[111] Frank Klassner, and Scott D Anderson. Lego mindstorms: Not just for k-12 anymore. *IEEE Robotics & Automation Magazine*, 10(2): 12–18, 2003.

[112] Ronald Kline. Cybernetics, automata studies, and the dartmouth conference on artificial intelligence. *IEEE Annals of the History of Computing*, 33(4): 5–16, 2010.

[113] Jelena Kocić, Nenad Jovičić, and Vujo Drndarević. An end-to-end deep neural network for autonomous driving designed for embedded automotive platforms. *Sensors*, 19(9): 2064, 2019.

[114] Teuvo Kohonen. Self-organizing neural projections. *Neural Networks*, 19(6-7): 723–733, 2006.

[115] Teuvo Kohonen. Essentials of the self-organizing map. *Neural Networks*, 37: 52–65, 2013.

[116] Teuvo Kohonen, and Timo Honkela. Kohonen network. *Scholarpedia*, 2(1): 1568, 2007.

[117] Ben Krause, Liang Lu, Iain Murray, and Steve Renals. Multiplicative lstm for sequence modelling. *arXiv preprint arXiv:1609.07959*, 2016.

[118] Alex Krizhevsky, Ilya Sutskever, and Geoffrey E Hinton. Imagenet classification with deep convolutional neural networks. *Advances in Neural Information Processing Systems*, 25: 1097–1105, 2012.

[119] Silvia Lameri, Federico Lombardi, Paolo Bestagini, Maurizio Lualdi, and Stefano Tubaro. Landmine detection from gpr data using convolutional neural networks. In *2017 25th European Signal Processing Conference (EUSIPCO)*, pp. 508–512. IEEE, 2017.

[120] Alan Lesgold, Susanne Lajoie, Marilyn Bunzo, and Gary Eggan. Sherlock: A coached practice environment for an electronics troubleshooting job. 1988.

[121] Qing Li, Weidong Cai, Xiaogang Wang, Yun Zhou, David Dagan Feng, and Mei Chen. Medical image classification with convolutional neural network. In *2014 13th International Conference on Control Automation Robotics & Vision (ICARCV)*, pp. 844–848. IEEE, 2014.

[122] Shang Li, Hongli Zhang, Lin Ye, Xiaoding Guo, and Binxing Fang. Mann: A multichannel attentive neural network for legal judgment prediction. *IEEE Access*, 7: 151144–151155, 2019.

[123] Yitong Li, Martin Min, Dinghan Shen, David Carlson, and Lawrence Carin. Video generation from text. In *Proceedings of the AAAI Conference on Artificial Intelligence*, volume 32, 2018.

[124] Daniele Liciotti, Michele Bernardini, Luca Romeo, and Emanuele Frontoni. A sequential deep learning application for recognising human activities in smart homes. *Neurocomputing*, 396: 501–513, 2020.

[125] Grace W Lindsay. Convolutional neural networks as a model of the visual system: Past, present, and future. *Journal of Cognitive Neuroscience*, 1–15, 2020.

[126] Richard Lippmann. An introduction to computing with neural nets. *IEEE Assp Magazine*, 4(2): 4–22, 1987.

[127] Geert Litjens, Thijs Kooi, Babak Ehteshami Bejnordi, Arnaud Arindra Adiyoso Setio, Francesco Ciompi, Mohsen Ghafoorian, Jeroen Awm Van Der Laak, Bram Van Ginneken, and Clara I Sánchez. A survey on deep learning in medical image analysis. *Medical Image Analysis*, 42: 60–88, 2017.

[128] Daniel Lopez-Martinez, Ke Peng, Arielle Lee, David Borsook, and Rosalind Picard. Pain detection with fnirs-measured brain signals: A personalized machine learning approach using the wavelet transform and bayesian hierarchical modeling with dirichlet process priors. In *2019 8th International Conference on Affective Computing and Intelligent Interaction Workshops and Demos (ACIIW)*, pp. 304–309. IEEE, 2019.

[129] Jiyun Luo, Sicong Zhang, and Hui Yang. Win-win search: Dual-agent stochastic game in session search. In *Proceedings of the 37th International ACM SIGIR Conference on Research & Development in Information Retrieval*, pp. 587–596, 2014.

[130] Fang Lv, Wei Wang, Yuliang Wei, Yunxiao Sun, Junheng Huang, and Bailing Wang. Detecting fraudulent bank account based on convolutional neural network with heterogeneous data. *Mathematical Problems in Engineering*, 2019.

[131] Pradeep Kumar Mallick, Seuc Ho Ryu, Sandeep Kumar Satapathy, Shruti Mishra, Gia Nhu Nguyen, and Prayag Tiwari. Brain mri image classification for cancer detection using deep wavelet autoencoder-based deep neural network. *IEEE Access*, 7: 46278–46287, 2019.

[132] Alessandro Massaro, Sergio Selicato, and Angelo Galiano. Predictive maintenance of bus fleet by intelligent smart electronic board implementing artificial intelligence. *IoT*, 1(2): 180–197, 2020.

[133] Warren S McCulloch, and Walter Pitts. A logical calculus of the ideas immanent in nervous activity. The Bulletin of Mathematical Biophysics, 5(4): 115–133, 1943.

[134] Niall McLaughlin, Jesus Martinez del Rincon, BooJoong Kang, Suleiman Yerima, Paul Miller, Sakir Sezer, Yeganeh Safaei, Erik Trickel, Ziming Zhao, Adam Doupé, and Gail Joon Ahn. Deep android malware detection. In *Proceedings of the Seventh ACM on Conference on Data and Application Security and Privacy*, pp. 301–308, 2017.

[135] Marvin L Minsky, and Seymour A Papert. Perceptrons: Expanded edition, 1988.

[136] Shariq Mobin and Joan Bruna. Voice conversion using convolutional neural networks. *arXiv preprint arXiv:1610.08927*, 2016.

[137] Elena Mocanu, Phuong H Nguyen, and Madeleine Gibescu. Deep learning for power system data analysis. In *Big Data Application in Power Systems*, pp. 125–158. Elsevier, 2018.

[138] Amin Hedayati Moghaddam, Moein Hedayati Moghaddam, and Morteza Esfandyari. Stock market index prediction using artificial neural network. *Journal of Economics, Finance and Administrative Science*, 21(41): 89–93, 2016.

[139] Md Shirajum Munir, Sarder Fakhrul Abedin, Md Golam Rabiul Alam, Do Hyeon Kim, and Choong Seon Hong. Rnn based energy demand prediction for smart home in smart-grid framework. pp. 437–439, 2017.

[140] Masaki Nakada, Han Wang, and Demetri Terzopoulos. Acfr: Active face recognition using convolutional neural networks. In *Proceedings of the IEEE Conference on Computer Vision and Pattern Recognition Workshops*, pp. 35–40, 2017.

[141] Yurii Nesterov. A method for unconstrained convex minimization problem with the rate of convergence o (1/k^ 2). In *Doklady an ussr*, volume 269, pp. 543–547, 1983.

[142] Lukáš Neumann, and Jiří Matas. Real-time scene text localization and recognition. In *2012 IEEE Conference on Computer Vision and Pattern Recognition*, pp. 3538–3545. IEEE, 2012.

[143] Johnatan S Oliveira, Gustavo B Souza, Anderson R Rocha, Flavio E Deus, and Aparecido N Marana. Cross-domain deep face matching for real banking security systems. In *2020 Seventh International Conference on eDemocracy & eGovernment (ICEDEG)*, pp. 21–28. IEEE, 2020.

[144] Yuki Onishi, Takeshi Yoshida, Hiroki Kurita, Takanori Fukao, Hiromu Arihara, and Ayako Iwai. An automated fruit harvesting robot by using deep learning. *Robomech Journal*, 6(1): 1–8, 2019.

[145] Yaakov Ophir, Refael Tikochinski, Christa SC Asterhan, Itay Sisso, and Roi Reichart. Deep neural networks detect suicide risk from textual facebook posts. *Scientific Reports*, 10(1): 1–10, 2020.

[146] Hasmik Osipyan, Somaiyeh Vedadi, and Adrian David Cheok. Machines as an assistants for humans' creativity: A conceptual model. In *IECON 2017-43rd Annual Conference of the IEEE Industrial Electronics Society*, pp. 3316–3321. IEEE, 2017.

[147] Emilio Parisotto, Abdel-rahman Mohamed, Rishabh Singh, Lihong Li, Dengyong Zhou, and Pushmeet Kohli. Neuro-symbolic program synthesis. *arXiv preprint arXiv:1611.01855*, 2016.

[148] David Paulius, and Yu Sun. A survey of knowledge representation in service robotics. *Robotics and Autonomous Systems*, 118: 13–30, 2019.

[149] Panagiotis Petsagkourakis, I Orson Sandoval, Eric Bradford, Dongda Zhang, and Ehecatl Antonio del Rio-Chanona. Reinforcement learning for batch-to-batch bioprocess optimisation. In *Computer Aided Chemical Engineering*, volume 46, pp. 919–924. Elsevier, 2019.

[150] Michael Phi. Illustrated guide to lstm's and gru's: A step by step explanation, 2018. *URL https://towardsdatascience. com/illustrated-guide-tolstms-and-gru-sa-step-by-step-explanation-44e9eb85bf21.*

[151] Harry A Pierson, and Michael S Gashler. Deep learning in robotics: A review of recent research. *Advanced Robotics*, 31(16): 821–835, 2017.

[152] Rimma Pivovarov, and Noémie Elhadad. Automated methods for the summarization of electronic health records. *Journal of the American Medical Informatics Association*, 22(5): 938–947, 2015.

[153] Dean Pomerleau. Alvinn: An autonomous land vehicle in a neural network. pp. 305–313. *In*: D.S. Touretzky (ed.). *Proceedings of Advances in Neural Information Processing Systems* 1. Morgan Kaufmann, December 1989.

[154] Dan Popa, Florin Pop, Cristina Serbanescu, and Aniello Castiglione. Deep learning model for home automation and energy reduction in a smart home environment platform. *Neural Computing and Applications*, 31(5): 1317–1337, 2019.

[155] Matti Pouke, and Jonna Häkkilä. Elderly healthcare monitoring using an avatar-based 3d virtual environment. *International Journal of Environmental Research and Public Health*, 10(12): 7283–7298, 2013.

[156] Ning Qian. On the momentum term in gradient descent learning algorithms. *Neural Networks*, 12(1): 145–151, 1999.

[157] Yanru Qu, Bohui Fang, Weinan Zhang, Ruiming Tang, Minzhe Niu, Huifeng Guo, Yong Yu, and Xiuqiang He. Product-based neural networks for user response prediction over multi-field categorical data. *ACM Transactions on Information Systems (TOIS)*, 37(1): 1–35, 2018.

[158] Alec Radford, Luke Metz, and Soumith Chintala. Unsupervised representation learning with deep convolutional generative adversarial networks. *arXiv preprint arXiv:1511.06434*, 2015.

[159] Siddhant Rai, Akshayanand Raut, Akash Savaliya, and Radha Shankarmani. Darwin: Convolutional neural network based intelligent health assistant. In *2018 Second International Conference on Electronics, Communication and Aerospace Technology (ICECA)*, pp. 1367–1371. IEEE, 2018.

[160] Vadlamani Ravi, Dadabada Pradeepkumar, and Kalyanmoy Deb. Financial time series prediction using hybrids of chaos theory, multi-layer perceptron and multi-objective evolutionary algorithms. *Swarm and Evolutionary Computation*, 36: 136–149, 2017.

[161] Asha RB, and Suresh Kumar KR. Credit card fraud detection using artificial neural network. *Global Transitions Proceedings*, 2(1): 35–41, 2021. 1st International Conference on Advances in Information, Computing and Trends in Data Engineering (AICDE-2020).

[162] Mittapalle Kiran Reddy, Paavo Alku, and Krothapalli Sreenivasa Rao. Detection of specific language impairment in children using glottal source features. *IEEE Access*, 8: 15273–15279, 2020.

[163] Russell Reed, and Robert J Marks II. *Neural Smithing: Supervised Learning in Feedforward Artificial Neural Networks*. Mit Press, 1999.

[164] Jing Ren, Mark Green, and Xishi Huang. From traditional to deep learning: Fault diagnosis for autonomous vehicles. In *Learning Control*, pp. 205–219. Elsevier, 2021.

[165] Adam Roegiest, Alexander K Hudek, and Anne McNulty. A dataset and an examination of identifying passages for due diligence. In *The 41st International ACM SIGIR Conference on Research & Development in Information Retrieval*, pp. 465–474, 2018.

[166] Jan Joris Roessingh, Armon Toubman, Joost van Oijen, Gerald Poppinga, Rikke Amilde LØvlid, Ming Hou, and Linus Luotsinen. Machine learning techniques for autonomous agents in military simulations—multum in parvo. In *2017 IEEE International Conference on Systems, Man, and Cybernetics (SMC)*, pp. 3445–3450. IEEE, 2017.

[167] Raúl Rojas. Kohonen networks. In *Neural Networks*, pp. 389–410. Springer, 1996.

[168] Lorenzo Rosasco, Ernesto De Vito, Andrea Caponnetto, Michele Piana, and Alessandro Verri. Are loss functions all the same? Neural Computation, 16(5): 1063–1076, 2004.

[169] Frank Rosenblatt. The perceptron: A probabilistic model for information storage and organization in the brain. *Psychological Review*, 65(6): 386–408, 1958.

[170] David Rotman. How technology is destroying jobs. *Technology Review*, 16(4): 28–35, 2013.

[171] Sebastian Ruder. An overview of gradient descent optimization algorithms. *arXiv preprint arXiv:1609.04747*, 2016.

[172] Andrei A Rusu, Neil C Rabinowitz, Guillaume Desjardins, Hubert Soyer, James Kirkpatrick, Koray Kavukcuoglu, Razvan Pascanu, and Raia

Hadsell. Progressive neural networks. *arXiv preprint arXiv:1606.04671*, 2016.

[173] Peyman Sabouri, and Hamid Gholam Hosseini. Lesion border detection using deep learning. In *2016 IEEE Congress on Evolutionary Computation (CEC)*, pp. 1416–1421. IEEE, 2016.

[174] Sumit Saha. A comprehensive guide to convolutional neural networks—the eli5 way, 15 december 2018. *URL: https://towardsdatascience. com/ a-comprehensive-guide-toconvolutionalneural-networks-the-eli5-way-3bd2b1164a53, 2020.*

[175] Ramadass Sathya, and Annamma Abraham. Comparison of supervised and unsupervised learning algorithms for pattern classification. *International Journal of Advanced Research in Artificial Intelligence*, 2(2): 34–38, 2013.

[176] Murat Hüsnü Sazlı. A brief review of feed-forward neural networks. 2006.

[177] Moritz Schwyzer, Daniela A Ferraro, Urs J Muehlematter, Alessandra Curioni-Fontecedro, Martin W Huellner, Gustav K Von Schulthess, Philipp A Kaufmann, Irene A Burger, and Michael Messerli. Automated detection of lung cancer at ultralow dose pet/ct by deep neural networks–initial results. *Lung Cancer*, 126: 170–173, 2018.

[178] Sungyong Seo, Jing Huang, Hao Yang, and Yan Liu. Interpretable convolutional neural networks with dual local and global attention for review rating prediction. In *Proceedings of the Eleventh ACM Conference on Recommender Systems*, pp. 297–305, 2017.

[179] Pierre Sermanet, and Yann LeCun. Traffic sign recognition with multiscale convolutional networks. In *The 2011 International Joint Conference on Neural Networks*, pp. 2809–2813. IEEE, 2011.

[180] Huma Shah, Kevin Warwick, Jordi Vallverdú, and Defeng Wu. Can machines talk? Comparison of eliza with modern dialogue systems. *Computers in Human Behavior*, 58: 278–295, 2016.

[181] K Gnana Sheela, and Subramaniam N Deepa. Review on methods to fix number of hidden neurons in neural networks. *Mathematical Problems in Engineering*, 2013.

[182] P Sibi, S Allwyn Jones, and P Siddarth. Analysis of different activation functions using back propagation neural networks. *Journal of Theoretical and Applied Information Technology*, 47(3): 1264–1268, 2013.

[183] Gyanendra Singh, Mahesh Pal, Yogender Yadav, and Tushar Singla. Deep neural network-based predictive modeling of road accidents. *Neural Computing and Applications*, pp. 1–10, 2020.

[184] Sneha Singhania, Nigel Fernandez, and Shrisha Rao. 3han: A deep neural network for fake news detection. In *International Conference on Neural Information Processing*, pp. 572–581. Springer, 2017.

[185] J Adeline Sneha, and Rekha Chakravarthi. Neural network based pest identification and control in cauliflower crop using sounds of pest. *International Journal of Innovative Technology and Exploring Engineering (IJITEE)*, 2019.

[186] Nitish Srivastava, Geoffrey Hinton, Alex Krizhevsky, Ilya Sutskever, and Ruslan Salakhutdinov. Dropout: A simple way to prevent neural networks from overfitting. *The Journal of Machine Learning Research*, 15(1): 1929–1958, 2014.

[187] Chao-Ton Su, Taho Yang, and Chir-Mour Ke. A neural-network approach for semiconductor wafer post-sawing inspection. *IEEE Transactions on Semiconductor Manufacturing*, 15(2): 260–266, 2002.

[188] Hung-Yue Suen, Kuo-En Hung, and Chien-Liang Lin. Intelligent video interview agent used to predict communication skill and perceived personality traits. *Human-centric Computing and Information Sciences*, 10(1): 1–12, 2020.

[189] Alagiah Suthakaran, and Saminda Premaratne. Detection of the affected area and classification of pests using convolutional neural networks from the leaf images. *International Journal of Computer Science Engineering (IJCSE)*, 2020.

[190] Ilya Sutskever, Oriol Vinyals, and Quoc V Le. Sequence to sequence learning with neural networks. In *Advances in Neural Information Processing Systems*, pp. 3104–3112, 2014.

[191] Keisuke Suzuki, Warrick Roseboom, David J Schwartzman, and Anil K Seth. Hallucination machine: Simulating altered perceptual phenomenology with a deep-dream virtual reality platform. In *ALIFE 2018: The 2018 Conference on Artificial Life*, pp. 111–112. MIT Press, 2018.

[192] Katia P Sycara. Negotiation planning: An AI approach. *European Journal of Operational Research*, 46(2): 216–234, 1990.

[193] Nian Chi Tay, Connie Tee, Thian Song Ong, and Pin Shen Teh. Abnormal behavior recognition using cnn-lstm with attention mechanism. In *2019 1st International Conference on Electrical, Control and Instrumentation Engineering (ICECIE)*, pp. 1–5. IEEE, 2019.

[194] Max Tegmark. *Life 3.0: Being Human in the Age of Artificial Intelligence*. Knopf, 2017.

[195] Lucas Theis, Wenzhe Shi, Andrew Cunningham, and Ferenc Huszár. Lossy image compression with compressive autoencoders. *arXiv preprint arXiv:1703.00395*, 2017.

[196] Ajith Thomas, and John Hedley. Fumebot: A deep convolutional neural network controlled robot. *Robotics*, 8(3): 62, 2019.

[197] Shuo-Chang Tsai, Cheng-Huan Chen, Yi-Tzone Shiao, Jin-Shuei Ciou, and Trong-Neng Wu. Precision education with statistical learning and deep learning: A case study in taiwan. *International Journal of Educational Technology in Higher Education*, 17: 1–13, 2020.

[198] Alan M Turing. Computing machinery and intelligence. In *Parsing the Turing Test*, pp. 23–65. Springer, 2009.

[199] Ilkay Ulusoy, and Christopher M Bishop. Generative versus discriminative methods for object recognition. In *2005 IEEE Computer Society Conference on Computer Vision and Pattern Recognition (CVPR'05)*, volume 2, pp. 258–265. IEEE, 2005.

[200] Ashwani Kumar Upadhyay, and Komal Khandelwal. Applying artificial intelligence: Implications for recruitment. *Strategic HR Review*, 2018.

[201] Fjodor Van Veen. Neural network zoo prequel: cells and layers. *Retried from https://www. asimovinstitute. org/author/fjodorvanveen*, 2017.

[202] Ivan Vasilev, Daniel Slater, Gianmario Spacagna, Peter Roelants, and Valentino Zocca. *Python Deep Learning: Exploring deep learning techniques and neural network architectures with PyTorch, Keras, and TensorFlow*. Packt Publishing Ltd., 2019.

[203] Izhar Wallach, Michael Dzamba, and Abraham Heifets. Atomnet: A deep convolutional neural network for bioactivity prediction in structure-based drug discovery. *arXiv preprint arXiv:1510.02855*, 2015.

[204] Fei-Yue Wang, Jun Jason Zhang, Xinhu Zheng, Xiao Wang, Yong Yuan, Xiaoxiao Dai, Jie Zhang, and Liuqing Yang. Where does alphago go: From church-turing thesis to alphago thesis and beyond. *IEEE/CAA Journal of Automatica Sinica*, 3(2): 113–120, 2016.

[205] Kai Wang, Boris Babenko, and Serge Belongie. End-to-end scene text recognition. In *2011 International Conference on Computer Vision*, pp. 1457–1464. IEEE, 2011.

[206] Weiyu Wang, and Keng Siau. Artificial intelligence, machine learning, automation, robotics, future of work and future of humanity: A review and research agenda. *Journal of Database Management (JDM)*, 30(1): 61–79, 2019.

[207] Yuxuan Wang, RJ Skerry-Ryan, Daisy Stanton, Yonghui Wu, Ron J Weiss, Navdeep Jaitly, Zongheng Yang, Ying Xiao, Zhifeng Chen, Samy Bengio, Quoc Le, Yannis Agiomyrgiannakis, Rob Clark, and Rif A Saurous. Tacotron: Towards end-to-end speech synthesis. *arXiv preprint arXiv:1703.10135*, 2017.

[208] Kevin Warwick. *Artificial Intelligence: The Basics*. Routledge, 2013.

[209] Joseph Weizenbaum. Eliza—a computer program for the study of natural language communication between man and machine. *Communications of the ACM*, 9(1): 36–45, 1966.

[210] Bernard Widrow, and Michael A Lehr. Perceptrons, adalines, and backpropagation. *Arbib*, 4: 719–724, 1995.

[211] David P Williams. Underwater target classification in synthetic aperture sonar imagery using deep convolutional neural networks. In *2016 23rd International Conference on Pattern Recognition (ICPR)*, pp. 2497–2502. IEEE, 2016.

[212] Benjamin Wohl, Barry Porter, and Sarah Clinch. Teaching computer science to 5–7 year-olds: An initial study with scratch, cubelets and unplugged computing. In *Proceedings of the Workshop in Primary and Secondary Computing Education*, pp. 55–60, 2015.

[213] Carole-Jean Wu, David Brooks, Kevin Chen, Douglas Chen, Sy Choudhury, Marat Dukhan, Kim Hazelwood, Eldad Isaac, Yangqing Jia, Bill Jia, Tommer Leyvand, Hao Lu, Yang Lu, Lin Qiao, Brandon Reagen, Joe Spisak, Fei Sun, Andrew Tulloch, Peter Vajda, Menlo Park, Xiaodong Wang, Yanghan Wang, Bram Wasti, Yiming Wu, Ran Xian, Sungjoo Yoo, and Peizhao Zhang. Machine learning at facebook: Understanding inference at the edge. In *2019 IEEE International Symposium on High Performance Computer Architecture (HPCA)*, pp. 331–344. IEEE, 2019.

[214] Qi Wu, Damien Teney, Peng Wang, Chunhua Shen, Anthony Dick, and Anton van den Hengel. Visual question answering: A survey of methods and datasets. *Computer Vision and Image Understanding*, 163: 21–40, 2017.

[215] Tianyong Wu, Jierui Liu, Zhenbo Xu, Chaorong Guo, Yanli Zhang, Jun Yan, and Jian Zhang. Light-weight, inter-procedural and callback-aware resource leak detection for android apps. *IEEE Transactions on Software Engineering*, 42(11): 1054–1076, 2016.

[216] Tingmin Wu, Shigang Liu, Jun Zhang, and Yang Xiang. Twitter spam detection based on deep learning. In *Proceedings of the Australasian Computer Science Week Multiconference*, pp. 1–8, 2017.

[217] Yi Wu, Jongwoo Lim, and Ming-Hsuan Yang. Online object tracking: A benchmark. In *Proceedings of the IEEE Conference on Computer Vision and Pattern Recognition*, pp. 2411–2418, 2013.

[218] Yonghui Wu, Mike Schuster, Zhifeng Chen, Quoc V Le, Mohammad Norouzi, Wolfgang Macherey, Maxim Krikun, Yuan Cao, Qin Gao, Klaus Macherey, Jeff Klingner, Apurva Shah, Melvin Johnson, Xiaobing Liu, Łukasz Kaiser, Stephan Gouws, Yoshikiyo Kato, Taku Kudo, Hideto Kazawa, Keith Stevens, George Kurian, Nishant Patil, Wei Wang, Cliff Young, Jason Smith, Jason Riesa, Alex Rudnick, Oriol Vinyals, Greg Corrado, Macduff Hughes, and Jeffrey Dean. Google's neural machine translation system: Bridging the gap between human and machine translation. *arXiv preprint arXiv:1609.08144*, 2016.

[219] Yang Xin, Lingshuang Kong, Zhi Liu, Yuling Chen, Yanmiao Li, Hongliang Zhu, Mingcheng Gao, Haixia Hou, and Chunhua Wang. Machine learning and deep learning methods for cybersecurity. *IEEE Access*, 6: 35365–35381, 2018.

[220] Kui Xu. *Anomaly Detection through System and Program Behavior Modeling*. PhD thesis, Virginia Tech, 2014.

[221] Hao Yang, Chenxi Liu, Meixin Zhu, Xuegang Ban, and Yinhai Wang. How fast you will drive? Predicting speed of customized paths by deep neural network. *IEEE Transactions on Intelligent Transportation Systems*, 2021.

[222] Keng-Chieh Yang, Conna Yang, Pei-Yao Chao, and Po-Hong Shih. Applying artificial neural network to predict semiconductor machine outliers. *Mathematical Problems in Engineering*, 2013.

[223] Bayya Yegnanarayana. *Artificial Neural Networks*. PHI Learning Pvt. Ltd., 2009.

[224] Wei Yu, Kuiyuan Yang, Yalong Bai, Tianjun Xiao, Hongxun Yao, and Yong Rui. Visualizing and comparing alexnet and vgg using deconvolutional layers. In *Proceedings of the 33rd International Conference on Machine Learning*, 2016.

[225] Xiaoyong Yuan. Phd forum: Deep learning-based real-time malware detection with multi-stage analysis. In *2017 IEEE International Conference on Smart Computing (SMARTCOMP)*, pp. 1–2. IEEE, 2017.

[226] Matthew D Zeiler. Adadelta: An adaptive learning rate method. *arXiv preprint arXiv:1212.5701*, 2012.

[227] Weili Zeng, Juan Li, Zhibin Quan, and Xiaobo Lu. A deep graph embedded lstm neural network approach for airport delay prediction. *Journal of Advanced Transportation*, 2021.

[228] Cheng Xiang Zhai. Towards a game-theoretic framework for text data retrieval. *IEEE Data Eng. Bull.*, 39(3): 51–62, 2016.

[229] Jing Zhang, Long He, Manoj Karkee, Qin Zhang, Xin Zhang, and Zongmei Gao. Branch detection for apple trees trained in fruiting wall architecture

using depth features and regions-convolutional neural network (r-cnn). *Computers and Electronics in Agriculture*, 155: 386–393, 2018.

[230] Liheng Zhang, Charu Aggarwal, and Guo-Jun Qi. Stock price prediction via discovering multi-frequency trading patterns. In *Proceedings of the 23rd ACM SIGKDD International Conference on Knowledge Discovery and Data Mining*, pp. 2141–2149, 2017.

[231] Shuai Zhang, Lina Yao, Aixin Sun, and Yi Tay. Deep learning based recommender system: A survey and new perspectives. *ACM Computing Surveys (CSUR)*, 52(1): 1–38, 2019.

[232] Alex Zhavoronkov, Yan A Ivanenkov, Alex Aliper, Mark S Veselov, Vladimir A Aladinskiy, Anastasiya V Aladinskaya, Victor A Terentiev, Daniil A Polykovskiy, Maksim D Kuznetsov, Arip Asadulaev, Yury Volkov, Artem Zholus, Rim R Shayakhmetov, Alexander Zhebrak, Lidiya I Minaeva, Bogdan A Zagribelnyy, Lennart H Lee, Richard Soll, David Madge, Li Xing, Tao Guo, and Alán Aspuru-Guzik. Deep learning enables rapid identification of potent ddr1 kinase inhibitors. *Nature Biotechnology*, 37(9): 1038–1040, 2019.

[233] Jing Zheng, Xiao Fu, and Guijun Zhang. Research on exchange rate forecasting based on deep belief network. *Neural Computing and Applications*, 31(1): 573–582, 2019.

[234] Wei Zhong, Huchuan Lu, and Ming-Hsuan Yang. Robust object tracking via sparsity-based collaborative model. In *2012 IEEE Conference on Computer Vision and Pattern Recognition*, pp. 1838–1845. IEEE, 2012.

[235] Lichun Zhou. Product advertising recommendation in e-commerce based on deep learning and distributed expression. *Electronic Commerce Research*, 20(2): 321–342, 2020.

[236] Yao Zhou, and Gary G Yen. Evolving deep neural networks for movie box-office revenues prediction. In *2018 IEEE Congress on Evolutionary Computation (CEC)*, pp. 1–8. IEEE, 2018.

[237] Zheyu Zhou. An intelligent teaching assistant system using deep learning technologies. In *Proceedings of the 2020 5th International Conference on Machine Learning Technologies*, pp. 18–22, 2020.

[238] Jun-Yan Zhu, Taesung Park, Phillip Isola, and Alexei A Efros. Unpaired image-to-image translation using cycle-consistent adversarial networks. In *Proceedings of the IEEE International Conference on Computer Vision*, pp. 2223–2232, 2017.

[239] Hans-Georg Zimmermann, Ralph Neuneier, and Ralph Grothmann. Active portfolio-management based on error correction neural networks. In *NIPS*, pp. 1465–1472, 2001.

[240] Martin Armstrong, and Andrew Zisserman. Robust object tracking. In *Proceedings of the Asian Conference on Computer Vision, Singapore*, pp. 5–8, 1995.

[241] Mohammad Zolghadr, Seyed Armin Akhavan Niaki, and STA Niaki. Modeling and forecasting US presidential election using learning algorithms. *Journal of Industrial Engineering International*, 14(3): 491–500, 2018.

Index

A

Abstraction layer 4
Accessible classrooms 71
Accuracy 6, 7, 15, 17–19, 53, 58, 60, 62–64, 67, 73–75, 79–81, 84, 92, 95, 99, 103, 107–109, 114
Acoustic detection mechanism 59
Activation function 12–14, 18, 44, 46, 49
Adaptive learning 28, 69, 70
Agriculture 57, 59, 61
AI applications 1, 69, 76, 77, 85, 104, 112, 113
AI technologies 61, 81, 85, 88, 107, 112, 116
Air transportation 104, 105
Algorithms 2, 4, 6, 7, 9, 11, 14, 17, 22, 25–28, 41, 42, 53, 58–65, 69, 71, 73, 76, 78–83, 86–88, 90–92, 94, 97, 100, 101, 103–105, 107, 113–115, 117
Analysis 33, 41, 43, 57, 62, 70, 71, 79, 83–85, 87, 88, 91, 95, 96, 100, 106, 107
Applications 1, 4, 5, 7–9, 33, 37, 41, 42, 44, 46, 53, 57, 58, 60–62, 65, 67, 69–72, 74–77, 79, 81, 82, 84, 85, 87, 90–95, 97, 99, 103, 104, 106–108, 112–114, 117, 118
Applied math 11
Artificial intelligence 1, 5
Audio synthesis 97, 98, 100
Autonomous systems 104
Autonomous weapons 80

B

Backward propagation 11, 14
Banking 62–65, 81
Batch normalization 22, 23, 43
Biases 3, 11, 14, 18, 19, 25, 54, 81, 97
Bidirectional Recurrent Neural Network 48
Binary classification 15, 89
Binary cross entropy 15
Biological neuron 4
Boltzmann machines 5, 6, 30, 88
Business model 79

C

Categorical cross entropy 15
Chain rule 14, 16
Challenges of NN 4
Chatbots 5, 74, 79, 87, 88
Classification 3, 4, 13, 15, 16, 24, 25, 30, 33, 53, 58, 63, 72, 74, 80, 81, 84–86, 89, 92, 93, 95–97, 106
Commonsense knowledge 114
Computer engineering 106, 109
Computer system 1
Computer vision 8, 32, 33, 74, 80, 87, 88, 91
Computing power 113
Connected transportation infrastructure 105
Connectionism 5, 6
Content analysis 71
Contract review 79

Convolutional neural network 5, 30, 33, 34, 37, 38
Crisis prediction 62, 65
Crop yield prediction 60, 61
Cross validation 20
Cybernetics 4, 5
Cybersecurity 83–85

D

Data availability 113, 114
Data protection 85
Data scaling 21
Data transfer and compression 100
Decision making 8, 62, 65, 76, 81, 114
Deconvolutional neural network 37
Deep Belief Network 6, 30, 52, 53
Deep blue machine 4
Deep Convolutional Inverse Graphics Network 37, 39
Deep learning 2, 11, 30, 57
Deep NN applications 106, 113, 114, 118
Deep Residual Network 42
Defense systems 80
Discriminative architecture 7
Disease prediction 76, 113
Dreaming experience 76
Driverless tractors 58, 61
Drug discovery 75, 76
Due diligence 77, 79

E

Early diagnosis 74
Early stopping point 21
Education 57, 66, 67, 69, 70, 114, 116
Education data mining 67
Election predictions 79
Entertainment 91, 97, 99, 114, 117
Ethics 85, 113
Exchange rate forecasting 65

F

Face recognition 60, 64, 81
Feature extraction 2, 84
Feed forward neural network 30
Finance 62, 64, 65, 114
Financial text mining 66
Forecasting 62, 64, 65, 67, 78, 108, 118
Forward propagation 11, 12, 23
Fraud detection 62, 63, 65, 113
Fraud transaction 63
Future of deep neural networks 114

G

Gated Recurrent Neural Networks 46
Generative Adversarial Networks 6, 29, 30, 50–52, 89
Generative modeling 7, 51

H

Health assistants 74–76
Healthcare 8, 57, 71, 74, 76, 106, 114, 116
Hidden layers 2, 14, 20, 24, 25, 48, 51, 54, 55, 67, 109
Hopfield Network 48–50
Huber loss function 15
Hybrid method 7
Hyperbolic tangent 13

I

Ideal customer detection 66
Image processing 30, 37, 58, 72, 73, 80, 91
Individual traffic planning 105
Industrial engineering 109
Industrial revolution 111, 112, 118
Industry 6, 8, 58, 62, 64, 65, 76, 77, 81, 84, 87, 88, 98–100, 103–107, 109, 114, 117
Insurance 88, 112, 114
Intelligent call routing system 91
Intelligent game opponents 100
Intelligent system 96
Intelligent tutoring systems 66
IoT devices 59, 60
Irrigation tools 61

K

Knowledge representation 1
Kohonen Network 40–42

L

Labeling 93, 113
Law enforcement 82, 83, 105

Learning 1–7, 11–14, 17, 20, 21, 23, 25–28, 30, 32, 40–44, 46, 47, 49–53, 55, 57, 63, 66, 67, 69–71, 74, 76, 81, 84, 94–99, 107, 113, 115, 117, 118
Legal 76–79, 113, 116, 117
Legal judgment 78, 79
Legal research 77, 79
Loan applications 62, 65, 67
Logistics 83, 85, 115
Long Short-Term Memory 43
Loss function 13–16

M

Machine learning 1, 3, 6, 32
Malware detection 84
Manufacturing 106, 107, 109, 111, 114, 115
Market demand prediction 57, 61, 62
Marketing 85, 86, 91
McCulloch-Pitts Neuron 4
Mean squared error 14, 61
Medical diagnostics 76
Meta-learning 113, 118
Military 69, 79, 80, 82–85, 117
Multiclass classification 15, 16

N

Narrow AI 7, 118
Natural language processing 33, 66
Network configuration 23
Network regularization 19
Neural network 1–3, 5, 6, 29–31, 33, 34, 37–40, 44, 46, 48, 54, 70, 106, 114
Neurons 2–4, 6, 11, 12, 24, 25, 29, 33, 41, 42, 48, 49, 64
Normalized initialization 18

O

Object detection 72, 80
Object extraction 72
Oil price prediction 62
Online customer services 91
Optimization 14, 17, 23, 25, 61, 64, 96, 97, 100, 103, 114
Optimizers 17, 25, 28
Overfitting 4, 19–21, 24, 43

P

Pain management 76
Partial derivatives 16, 27
Pattern detection 60
Perceptron 4, 29, 48–50, 62
Personalized learning 69, 71
Politics 76–79
Portfolio management 64, 66
Preactivation 11, 12
Precision farming 61
Prediction 14, 19, 43, 44, 46, 57, 60–62, 64, 65, 67, 68, 72, 76, 77, 79, 82, 87, 90, 92, 95, 98, 99, 103, 105, 107–109, 113, 117, 118
Predictive analytics 70, 85, 86
Predictive fleet maintenance 105
Preventive care 71, 76
Profitable import and export 60
Public safety 81, 85

R

Reasoning 1, 52, 101, 114
Recognition systems 1
Rectified linear unit activation function 13
Recurrent Neural Network 30, 39, 40, 46, 48, 54
Remote treatment 76
Residual learning 23
Risk analysis 79
Risk management 66, 83, 108
Robots 58, 59, 66–68, 79, 80, 86–88, 98–101, 103, 115

S

Search optimization 100
Security 64, 65, 72, 79, 81, 83–85, 115, 117
Self-driving cars 8, 101
Self-learning 2
Self-organizing map 40, 62
Sensor 59, 75, 76, 81, 87, 91, 93, 99, 101–103
Sentiment analysis 87, 95, 96, 100
Service 6, 80, 83, 85–91, 101, 102, 107, 108, 115, 116
Sigmoid 12, 13, 15, 45, 46, 55
Simulation and training 82, 85

Smart homes and grids 108, 109
Smart recruitment 91
Smart spraying 61
Social media 78, 79, 83, 91, 92, 94, 99, 100
Social media listening 91, 100
Speech recognition 32, 43, 44, 53, 86, 88, 99, 114
Speech therapy 74
Stochastic gradient descent 5, 26
Stock market 64, 65
Strong AI 7
Super AI 7, 8
Supervised learning 2, 51, 53, 67, 98, 115
Support vector machines 5
Synapses 3

T

Tagging 100
Target detection 80, 85
Teaching assistants 69, 71
Technology 1, 7, 59–61, 71, 78, 84, 101, 103, 107, 111, 112, 115
Text recognition 74
Thinking machine 5
Trading performance 64
Trading strategies 64, 66
Traffic management and smart roads 105
Training 3, 4, 6, 14, 15, 18–23, 25, 26, 30, 32, 42, 43, 45, 46, 48, 50–52, 55, 67, 69, 70, 76, 82, 84, 85, 89, 93, 96, 114, 115
Training parameter 4
Translators 69
Transportation 83, 100–105, 114, 116

U

Underfitting 19, 20, 24
Unsupervised learning 2, 6, 7, 41, 51, 53, 63, 94

V

Vanishing gradient 13, 14, 23, 43, 45, 46, 90
Virtual adviser 79
Virtual assistant 1, 6, 71, 86–88, 91
Visual analysis 100
Voice conversion system 70

W

Warfare systems 85
Weighted sum 12
Weights 12, 17–20, 39, 41, 48, 49, 54